An Engineering Da

# An Engineering Data Book

*Edited by*

**JR Calvert and RA Farrar**

palgrave

First published 1999 by
PALGRAVE
Houndmills, Basingstoke, Hampshire RG21 6XS and
175 Fifth Avenue, New York, N. Y. 10010
Companies and representatives throughout the world

PALGRAVE is the new global academic imprint of
St. Martin's Press LLC Scholarly and Reference Division and
Palgrave Publishers Ltd (formerly Macmillan Press Ltd).

ISBN 0–333–51661–3

This book is printed on paper suitable for recycling and made from fully managed and sustained forest sources.

A catalogue record for this book is available from the British Library.

10 9 8 7 6 5 4 3
09 08 07 06 05 04 03 02 01

Printed and bound in Great Britain by
Creative Print & Design (Wales), Ebbw Vale

# Contents

# Preface

Learning formulae is the bugbear of the student, and many avoidable errors, both in education and professional practice, have been made through the incorrect recall of formulae or numbers. This booklet aims to provide a ready reference for commonly required formulae and data, for use in coursework and examinations (where permitted) and in professional practice. It is not a textbook – the user is expected to know when to use a particular formula, and just as important, when not to use it.

The units used are SI, or multiples, with conversion factors from other systems provided. The symbols are generally those in common use in particular specialities, except where this would cause confusion in context in a particular section of the book. A complete list of symbols, with meaning and units, is included.

This edition is an extended and revised version of that edited by AJ Munday and RA Farrar, published in 1979 in collaboration with The Macmillan Press Ltd, itself based on an internal publication of the Department of Mechanical Engineering, University of Southampton.

Data have been collected from a large variety of sources, and the editors are grateful to many colleagues, past and present, for their contributions and suggestions.

The editors have made every effort to ensure the accuracy of this data,but cannot guarantee it. No responsibility can be taken for the consequences of any errors which may remain. Any person finding an error is asked to inform the publishers.

# 1. Symbols And Units

## 1.1 *Symbols*

**The Greek Alphabet**

| | | | | | | | | | | | |
|---|---|---|---|---|---|---|---|---|---|---|---|
| Α | α | alpha | Η | η | eta | Ν | ν | nu | Τ | τ | tau |
| Β | β | beta | Θ | θ | theta | Ξ | ξ | xi | Υ,ϒ | υ | upsilon |
| Γ | γ | gamma | Ι | ι | iota | Ο | ο | omicron | Φ | ϕ | phi |
| Δ | δ | delta | Κ | κ | kappa | Π | π | pi | Χ | χ | chi |
| Ε | ε,ϵ | epsilon | Λ | λ | lambda | Ρ | ρ | rho | Ψ | ψ | psi |
| Ζ | ζ | zeta | Μ | μ | mu | Σ | σ | sigma | Ω | ω | omega |

**Mathematical Symbols**

| | | | |
|---|---|---|---|
| L[ ] | Laplace transform | $\underline{a}$ | Vector |
| $\Delta$ | Defined as | $\underline{\hat{a}}$ | Unit vector |
| $\Sigma$ | Repeated summation | $\perp$ | 'at right angles to' |
| $\Pi$ | Repeated multiplication | $\bullet$ | Scalar (dot) product |
| $\partial$ | Partial differentiation | $\times$, $\wedge$ | Vector (cross) product |
| $\lvert a \rvert$ | Modulus | Re( ) | Real part of complex number |
| $\nabla$ | Laplacian operator: Del, Nabla | Im( ) | Imaginary part of complex number |

**Decimal Prefixes**

| Symbol | Prefix | Multiplier | Symbol | Prefix | Multiplier |
|---|---|---|---|---|---|
| T | tera | $10^{12}$ | c | centi | $10^{-2}$ |
| G | giga | $10^{9}$ | m | milli | $10^{-3}$ |
| M | mega | $10^{6}$ | μ | micro | $10^{-6}$ |
| k | kilo | $10^{3}$ | n | nano | $10^{-9}$ |
| h | hecto | $10^{2}$ | p | pico | $10^{-12}$ |
| da | deca | 10 | f | femto | $10^{-15}$ |
| d | deci | $10^{-1}$ | a | atto | $10^{-18}$ |

## 1.2 ***SI Units***

**Basic Units**

| Quantity | Unit Name | Unit Symbol |
|---|---|---|
| Length | metre | m |
| Mass | kilogramme | kg |
| Time | second | s |
| Electric Current | ampere | A |
| Thermodynamic Temperature | kelvin | K |
| Luminous Intensity | candela | cd |

**Supplementary and Derived Units**

| Quantity | Unit Name | Unit Symbol | Equivalent Units | Dimensions MLT |
|---|---|---|---|---|
| Plane Angle | radian | rad | - | - |
| Solid Angle | steradian | sr | - | - |
| Force | Newton | N | kg m/s$^2$ | $MLT^{-2}$ |
| Work, Energy, Heat | Joule | J | N m | $ML^2T^{-2}$ |
| Power | Watt | W | J/s, Nm/s | $ML^2T^{-3}$ |
| Frequency | Hertz | Hz | $s^{-1}$ | $T^{-1}$ |
| Dynamic Viscosity | - | Ns/m$^2$ | kg/m s | $ML^{-1}T^{-1}$ |
| Kinematic Viscosity | - | m$^2$/s | - | $L^2T^{-1}$ |
| Pressure | Pascal | Pa | N/m$^2$ | $ML^{-1}T^{-2}$ |

***Electrical and Magnetic***

| | | | |
|---|---|---|---|
| Potential | Volt | V | W/A, Nm/As |
| Resistance | Ohm | Ω | V/A |
| Charge | Coulomb | C | A s |
| Current | Ampère | A | C/s |
| Capacitance | Farad | F | A s/V |
| Electric Field Strength | - | V/m | - |
| Electric Flux Density | - | C/m$^2$ | - |
| Magnetic Flux | Weber | Wb | V s, Nm/A |
| Inductance | Henry | H | V s/A, Nm/A$^2$ |
| Magnetic Field Strength | - | A/m | |
| Magnetic Flux Density | Tesla | T | Wb/m$^2$, N/(Am) |

***Light***

| | | | |
|---|---|---|---|
| Luminous Flux | lumen | lm | cd sr |
| Luminance | nit | cd/m$^2$ | - |
| Illumination | lux | lx | lm/m$^2$ |

## 1.3 *Conversion Factors for some other units into SI Units*

**Length**

| | | | | | | |
|---|---|---|---|---|---|---|
| 1 Å (Ångstrom) | = | $10^{-10}$ m | | 1 $\mu$ (micron) | = | 1 $\mu$m |
| 1 in (inch) | = | 25.4 mm (exactly) | | 1 $\mu$in | = | 0.0254 $\mu$m |
| 1 thou | = | 1 mil | = | 0.001 in | = | 25.4 $\mu$m |
| 1 ft (foot) | = | 0.3048 m | | 1 yd (yard) | = | 0.914 m |
| 1 mi (mile) | = | 5280 ft | = | 1.609 km | | |

**Area**

| | | | | |
|---|---|---|---|---|
| 1 acre | = | 0.4047 ha (hectare) | = | 4 047 $m^2$ |

**Volume**

| | | | | | | |
|---|---|---|---|---|---|---|
| 1 $in^3$ | = | 16.39 $cm^3$ | | 1 $ft^3$ | = | 0.02832 $m^3$ |
| 1 l (litre) | = | 1 $dm^3$ | = | $10^{-3}$ $m^3$ | | |
| 1 cc | = | 1 $cm^3$ | = | 1 ml | = | $10^{-6}$ $m^3$ |
| 1 gal (imperial gallon) | = | 0.1605 $ft^3$ | = | 4.546 l | = | 4546 $cm^3$ |
| 1 US gal | = | 0.1337 $ft^3$ | = | 3.785 l | = | 3.785 $cm^3$ |

**Mass**

| | | | | |
|---|---|---|---|---|
| 1 tonne (metric ton) | = | 1 Mg | = | 1000 kg |
| 1 lb (pound) | = | 0.4536 kg | | |
| 1 slug | = | 32.17 lb | = | 14.59 kg |
| 1 ton | = | 2240 lb | = | 1016 kg |

**Density**

| | | | | |
|---|---|---|---|---|
| 1 lb/$in^3$ | = | 27.68 g/$cm^3$ | = | 27 680 kg/$m^3$ |
| 1 lb/$ft^3$ | = | 16.02 kg/$m^3$ | | |
| 1 slug/$ft^3$ | = | 515.4 kg/$m^3$ | | |

**Velocity, angular velocity**

| | | | | | | |
|---|---|---|---|---|---|---|
| 1 mile/h | = | 1.467 ft/s | = | 1.609 km/h | = | 0.447 m/s |
| 1 knot | = | 1.689 ft/s | = | 1.853 km/h | = | 0.514 m/s |
| 1 rev/s | = | 6.283 rad/s | | 1000 rev/min | = | 104.7rad/s |

**Flow rate**

| | | | | | | |
|---|---|---|---|---|---|---|
| 1 $ft^3$/s | = | 1 cusec | = | 28.32 l/s | = | 0.02832 $m^3$/s |
| 1 cufm (cubic foot per minute) | = | 28.32 l/min | = | 1.699 $m^3$/h | | |
| 1 gal/min | = | 0.7577 l/s | = | $7.577 \times 10^{-5}$ $m^3$/s | | |

**Force, Weight**

| | | | | | | |
|---|---|---|---|---|---|---|
| 1 dyne | = | $10^{-5}$N | | 1 tonf (ton force) | = | 9964 N |
| 1 pdl (poundal) | = | 0.1383 N | | | | |
| 1 lbf (pound force) | = | 32.17 pdl | = | 4.448 N | | |
| 1 kp (kilopond) | = | 1 kgf | = | 2.205 lbf | = | 9.807 N |

**Torque**

| | | | | |
|---|---|---|---|---|
| 1 lbf in | = | 11.3 N cm | = | 0.113 Nm |
| 1 lbf ft | = | 1.356 Nm | | |
| 1 tonf ft | = | 3037 Nm | | |

**Stiffness**

| | | | | | | |
|---|---|---|---|---|---|---|
| 1 lbf/μ in | = | 175 N/μm | | 1 lbf/in | = | 17.5 N/m |

**Energy, Work, Heat**

| | | | | | | |
|---|---|---|---|---|---|---|
| 1 eV (electron volt) | = | $1.602 \times 10^{-19}$ J | | 1 ft lbf | = | 1.356 J |
| 1 kW h | = | 3.6 MJ | | 1 hp h | = | 2.685 MJ |
| 1 cal (calorie) | = | 4.187 J | | | | |
| 1 Btu (British thermal unit) | = | 778.2 ft lbf | = | 252 cal | = | 1055 J |
| 1 Chu (Centigrade heat unit) | = | 1.8 Btu | = | 1899 J | | |

**Power, Heat Flow**

| | | | | | | |
|---|---|---|---|---|---|---|
| 1 ft lbf/s | = | 1.356 W | | | | |
| 1 hp (horsepower) | = | 550 ft lbf/s | = | 33 000 ft lbf/min | = | 0.7457 kW |
| 1 ch (metric horsepower) | = | 1 PS | = | 0.7355 kW | | |

**Pressure, Vacuum, Stress**

| | | | | |
|---|---|---|---|---|
| 1 lbf/in$^2$ | = | 0.07031 kgf/cm$^2$ | = | 6895 N/m$^2$ |
| 1 lbf/ft$^2$ | = | 47.88 N/m$^2$ | | |
| 1 tonf/in$^2$ | = | 157.5 kgf/cm$^2$ | = | 15.44 MN/m$^2$ |
| 1 kgf/cm$^2$ | = | 0.09807 MN/m$^2$ | = | 0.9807 bar |
| 1 dyne/cm$^2$ | = | 0.1 Pa | = | 0.1 N/m$^2$ |
| 1 kgf/mm$^2$ | = | 9.807 MN/m$^2$ | = | 98.07 bar |
| 1 bar | = | 0.1 MPa | = | $10^5$ N/m$^2$ |
| | = | 14.50 lbf/in$^2$ | | |
| 1 atm (international atmosphere) | = | 14.70 lbf/in$^2$ | = | 101.3 kPa |
| | | | = | 1.013 bar |

**Stress Intensity**

| | | |
|---|---|---|
| 1 ksi $\sqrt{\text{in}}$ (kilo-pound per square inch) | = | 1.10 MNm$^{-3/2}$ |

**Head**

| | | | | |
|---|---|---|---|---|
| 1 ft $H_2O$ | ≈ | 62.43 lbf/ft$^2$ | = | 2989 Pa |
| 1 cm $H_2O$ | ≈ | 9.81 Pa | | |
| 1 in Hg | = | 13.6 in $H_2O$ | ≈ | 3386 Pa |
| 1 mm Hg | = | 1 torr | ≈ | 133.3 N/m$^2$ |
| 1 int atm | = | 1.013 bar | ≈ | 10.34 m water |
| | | | ≈ | 760 mm Hg |

**Dynamic Viscosity**

| | | | | |
|---|---|---|---|---|
| 1 P (poise) | = | 1 g/(cm s) | = | 0.1 Ns/m$^2$ |
| 1 cP | = | $10^{-3}$ kg/(m s) | = | 1 mN s/m$^2$ |
| 1 kgf s/m$^2$ | = | 9.807 N s/m$^2$ | | |

| | | | | | | |
|---|---|---|---|---|---|---|
| 1 lb/ft h | = | 0.4132 mN s/m$^2$ | | | | |
| 1 slug/ft s | = | 1 lbf s/ft$^2$ | = | 47.88 N s/m$^2$ | | |
| 1 Reyn | = | 1 lbf s/in$^2$ | = | 6895 N s/m$^2$ | | |

**Kinematic Viscosity**

| | | | | |
|---|---|---|---|---|
| 1 ft$^2$/s | = | 0.09290 m$^2$/s | | |
| 1 in$^2$/s | = | 645.2 mm$^2$/s | | |
| 1 St (Stokes) | = | 1 cm$^2$/s | = | $10^{-4}$ m$^2$/s |
| 1 cSt | = | 1 mm$^2$/s | = | $10^{-6}$ m$^2$/s |

**Temperature**

| | | | | | | |
|---|---|---|---|---|---|---|
| $T$ °C | = | $\frac{5}{9}(T\,°\text{F} - 32)$ | | $T$ °F | = | $\frac{9}{5}(T\,°\text{C}) + 32$ |
| $(\Delta T)$°C | = | $(\Delta T)$K | = | $\frac{5}{9}(\Delta T)$°F | = | $\frac{5}{9}(\Delta T)$°R |
| $T$ K | = | $\frac{5}{9}(T\,°\text{F} + 459.67)$ | = | $\frac{5}{9}T$ °R | = | $T$ °C + 273.15 |

**Specific Energy (internal energy, enthalpy, latent heat, calorific value)**

| | | |
|---|---|---|
| 1 Btu/lb | = | 2.326 kJ/kg |
| 1 ft lbf/lb | = | 2.989 J/kg |

**Specific Entropy, Specific Heat**

| | | |
|---|---|---|
| 1 Btu/(lb °R) | = | 4.187 kJ/(kg K) |

**Thermal Conductivity**

| | | | | |
|---|---|---|---|---|
| 1 Btu/(ft h °R) | = | 1.731 J/(m s °C) | = | 1.731 W/(m K) |
| 1 cal/(cm s K) | = | 418.7 J/(m s °C) | = | 418.7 W/(m K) |

**Light**

| | | | | |
|---|---|---|---|---|
| 1 lm/ft$^2$ | = | 1 foot-candle | = | 10.763 lx |
| 1 cd/ft$^2$ | = | 10.764 cd/m$^2$ | | |
| 1 foot-Lambert | = | 3.426 cd/m$^2$ | | |

## 2. Physical Constants

| | | | |
|---|---|---|---|
| Avogadro's number | $N$ | = | $6.023\times10^{26}$/(kg mol) |
| Bohr magneton | $\beta$ | = | $9.274\times10^{-24}$ A m$^2$ |
| Boltzmann's constant | $k$ | = | $1.381\times10^{-23}$ J/K |
| Stefan-Boltzmann constant | $\sigma$ | = | $5.67\times10^{-8}$ W/(m$^2$K$^4$) |
| electron charge | $e$ | = | $1.602\times10^{-19}$ C |
| electronic rest mass | $m_e$ | = | $9.109\times10^{-31}$ kg |
| electronic charge to mass ratio | $e/m_e$ | = | $1.759\times10^{11}$ C/kg |
| Faraday constant | $F$ | = | $9.65\times10^{7}$ C/(kg mol) |
| permeability of free space | $\mu_0$ | = | $4\pi\times10^{-7}$ H/m |
| permittivity of free space | $\varepsilon_0$ | = | $8.854\times10^{-12}$ F/m |
| characteristic impedance of free space | $Z_0$ | = | $(\mu_0/\varepsilon_0)^{\frac{1}{2}} = 120\pi\ \Omega$ |
| Planck's constant | $h$ | = | $6.626\times10^{-34}$ J s |
| proton mass | $m_p$ | = | $1.672\times10^{-27}$ kg |
| proton to electron mass ratio | $m_p/m_e$ | = | 1836.1 |
| standard gravitational intensity | $g$ | = | 9.80665 N/kg |
| (acceleration) | | = | 9.80665 m/s$^2$ |
| unified atomic mass constant | $u$ | = | $1.661\times10^{-27}$ kg |
| universal constant of gravitation | $G$ | = | $6.67\times10^{-11}$ N m$^2$/kg$^2$ |
| universal gas constant | $R_0$ | = | 8.314 kJ/(kg mol K) |
| velocity of light in vacuo | $c$ | = | $2.9979\times10^{8}$ m/s |
| volume of ideal gas at 1 atm, 0°C (Standard Temp. and Pressure – STP) | | = | 22.41 m$^3$/(kg mol) |

# 3 Analysis

## 3.1 *Vector Algebra*

$$\hat{\underline{a}} = \frac{\underline{a}}{|\underline{a}|}$$

$$\underline{a} = a_1\underline{i} + a_2\underline{j} + a_3\underline{k} \equiv (a_1, a_2, a_3)$$

$$a \equiv |\underline{a}| = \sqrt{a_1^2 + a_2^2 + a_3^2}$$

$$\underline{a} + \underline{b} = (a_1 + b_1, a_2 + b_2, a_3 + b_3)$$

**Scalar (dot) Product:**

$$\underline{a} \bullet \underline{b} = (a_1 b_1 + a_2 b_2 + a_3 b_3) = a\, b \cos\theta$$

$$\underline{a} \bullet \underline{a} = a^2$$

**Vector (cross) Product:**

$$\underline{a} \times \underline{b} = \begin{vmatrix} \underline{i} & \underline{j} & \underline{k} \\ a_1 & a_2 & a_3 \\ b_1 & b_2 & b_3 \end{vmatrix} = ab \sin\theta\, \hat{\underline{n}}, \quad \text{where } \hat{\underline{n}} \perp \underline{a},\ \hat{\underline{n}} \perp \underline{b}$$

$$\underline{a} \times \underline{a} = 0$$

Moment: $\underline{M} = \underline{r} \times \underline{F}$

Velocity: $\underline{V} = \underline{\omega} \times \underline{r}$

**Triple Scalar Product:**

$$[\underline{a}\,\underline{b}\,\underline{c}] = \underline{a} \bullet \underline{b} \times \underline{c} = \underline{a} \times \underline{b} \bullet \underline{c} = \begin{vmatrix} a_1 & a_2 & a_3 \\ b_1 & b_2 & b_3 \\ c_1 & c_2 & c_3 \end{vmatrix}$$

**Triple Vector Product:**

$$\underline{a} \times (\underline{b} \times \underline{c}) = (\underline{a} \bullet \underline{c})\underline{b} - (\underline{a} \bullet \underline{b})\underline{c}$$

$$(\underline{a} \times \underline{b}) \times \underline{c} = (\underline{a} \bullet \underline{c})\underline{b} - (\underline{b} \bullet \underline{c})\underline{a}$$

**Differentiation of Vectors:**

$$\frac{d}{dt}(\underline{a} + \underline{b}) = \frac{d\underline{a}}{dt} + \frac{d\underline{b}}{dt} \qquad \frac{d}{dt}(f\,\underline{a}) = \frac{df}{dt}\underline{a} + f\frac{d\underline{a}}{dt}$$

$$\frac{d}{dt}(\underline{a} \bullet \underline{b}) = \underline{a} \bullet \frac{d\underline{b}}{dt} + \frac{d\underline{a}}{dt} \bullet \underline{b} \qquad \frac{d}{dt}(\underline{a} \times \underline{b}) = \underline{a} \times \frac{d\underline{b}}{dt} + \frac{d\underline{a}}{dt} \times \underline{b}$$

$$\frac{d}{dt}(\underline{a}\bullet\underline{b}\times\underline{c}) = \frac{d\underline{a}}{dt}\bullet\underline{b}\times\underline{c} + \underline{a}\bullet\frac{d\underline{b}}{dt}\times\underline{c} + \underline{a}\bullet\underline{b}\times\frac{d\underline{c}}{dt}$$

**Equation of a Straight Line:**

$\underline{r} = \underline{a} + \lambda\underline{b}$

**Equation of a Plane:**

$(\underline{r} - \underline{a})\bullet\underline{c} = 0$ or $\underline{r}\bullet\underline{c} = \beta$

**Gradient of a scalar:**

$$\text{grad } V = \nabla V = \underline{i}\frac{\partial V}{\partial x} + \underline{j}\frac{\partial V}{\partial y} + \underline{k}\frac{\partial V}{\partial z} \quad \text{(Cartesian)}$$

$$= \underline{u}_r\frac{\partial V}{\partial r} + \underline{u}_\phi\frac{1}{r}\frac{\partial V}{\partial \phi} + \underline{u}_z\frac{\partial V}{\partial z} \quad \text{(Cylindrical)}$$

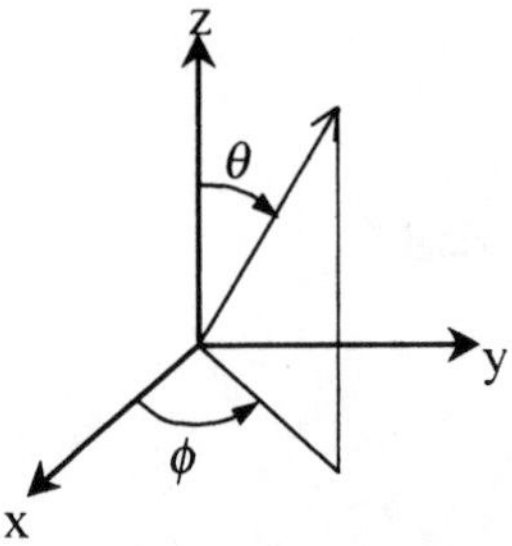


where:

$\underline{u}_r = \underline{i}\cos\phi + \underline{j}\sin\phi$

$\underline{u}_\phi = -\underline{i}\sin\phi + \underline{j}\cos\phi$

$\underline{u}_z = \underline{k}$

$$= \underline{u}_r\frac{\partial V}{\partial r} + \frac{\underline{u}_\theta}{r}\frac{\partial V}{\partial \theta} + \frac{\underline{u}_\phi}{r\sin\theta}\frac{\partial V}{\partial \phi} \quad \text{(Spherical)}$$

where:

$\underline{u}_r = \underline{i}\cos\phi\sin\theta + \underline{j}\sin\phi\sin\theta + \underline{k}\cos\theta$

$\underline{u}_\theta = \underline{i}\cos\phi\cos\theta + \underline{j}\sin\phi\cos\theta - \underline{k}\sin\theta$

$\underline{u}_\phi = -\underline{i}\sin\phi + \underline{j}\cos\phi$

**Divergence of a vector:**

$$\text{div F} = \nabla\bullet\underline{F} = \frac{\partial F_x}{\partial x} + \frac{\partial F_y}{\partial y} + \frac{\partial F_z}{\partial z} \quad \text{(Cartesian)}$$

$$= \frac{1}{r}\frac{\partial}{\partial r}(rF_r) + \frac{1}{r}\frac{\partial F_\phi}{\partial \phi} + \frac{\partial F_z}{\partial z} \quad \text{(Cylindrical}$$

$$= \frac{1}{r^2}\frac{\partial}{\partial r}(r^2 F_r) + \frac{1}{r\sin\theta}\frac{\partial}{\partial \theta}(F_\theta\sin\theta) + \frac{1}{r\sin\theta}\frac{\partial F_\phi}{\partial \phi} \quad \text{(Spherical)}$$

**Curl of a vector:**

$$\text{curl}\,\underline{F} = \text{rot}\,\underline{F} = \nabla \times \underline{F} = \underline{i}\left(\frac{\partial F_z}{\partial y} - \frac{\partial F_y}{\partial z}\right) + \underline{j}\left(\frac{\partial F_x}{\partial z} - \frac{\partial F_z}{\partial x}\right) + \underline{k}\left(\frac{\partial F_y}{\partial x} - \frac{\partial F_x}{\partial y}\right)$$

$$= \begin{vmatrix} \underline{i} & \underline{j} & \underline{k} \\ \frac{\partial}{\partial x} & \frac{\partial}{\partial y} & \frac{\partial}{\partial z} \\ F_x & F_y & F_z \end{vmatrix} \quad \text{(Cartesian)}$$

$$= \frac{1}{r}\begin{vmatrix} \underline{u}_r & r\underline{u}_\phi & \underline{u}_z \\ \frac{\partial}{\partial r} & \frac{\partial}{\partial \phi} & \frac{\partial}{\partial z} \\ F_r & rF_\phi & F_z \end{vmatrix} \quad \text{(Cylindrical)}$$

$$= \frac{1}{r^2 \sin\theta}\begin{vmatrix} \underline{u}_r & r\underline{u}_\theta & r\sin\theta\,\underline{u}_\phi \\ \frac{\partial}{\partial r} & \frac{\partial}{\partial \theta} & \frac{\partial}{\partial \phi} \\ F_r & rF_\theta & r\sin\theta\,F_\phi \end{vmatrix} \quad \text{(Spherical)}$$

**Laplace's Equation:**

$$\nabla \bullet \nabla V = \nabla^2 V = \frac{\partial^2 V}{\partial x^2} + \frac{\partial^2 V}{\partial y^2} + \frac{\partial^2 V}{\partial z^2} \quad \text{(Cartesian)}$$

$$= \frac{1}{r}\frac{\partial}{\partial r}\left(r\frac{\partial V}{\partial r}\right) + \frac{1}{r^2}\frac{\partial^2 V}{\partial \phi^2} + \frac{\partial^2 V}{\partial z^2} \quad \text{(Cylindrical)}$$

$$= \frac{1}{r^2}\frac{\partial}{\partial r}\left(r^2\frac{\partial V}{\partial r}\right) + \frac{1}{r^2\sin\theta}\frac{\partial}{\partial \theta}\left(\sin\theta\frac{\partial V}{\partial \theta}\right) + \frac{1}{r^2\sin^2\theta}\frac{\partial^2 V}{\partial \phi^2} \quad \text{(Spherical)}$$

**Space Curves:**

$\underline{v} = \underline{u}\dfrac{ds}{dt}$ $\qquad s$ = arc length, $\underline{u}$ = unit tangent

$\underline{a} = \dfrac{v^2}{\rho}\underline{n} + \dfrac{dv}{dt}\underline{u}$ $\qquad \underline{n}$ = unit 'inward' normal

$\dfrac{d\underline{u}}{ds} = \dfrac{1}{\rho}\underline{n}$ $\qquad \rho$ = radius of curvature

$\underline{b} = \underline{u} \times \underline{n}$ $\qquad \underline{b}$ = binormal vector

$\dfrac{d\underline{b}}{ds} = \dfrac{1}{\tau}\underline{n}, \quad \dfrac{d\underline{n}}{ds} = \dfrac{1}{\tau}\underline{b} - \dfrac{1}{\rho}\underline{u},$ $\qquad \dfrac{1}{\tau}$ = torsion

**Identities:**

$$\nabla \bullet \phi \underline{u} = \phi \nabla \bullet \underline{u} + \underline{u} \bullet \nabla \phi$$

$$\nabla \times \phi \underline{u} = \phi \nabla \times \underline{u} + \nabla \phi \times \underline{u}$$

$$\nabla \bullet \underline{u} \times \underline{v} = \underline{v} \bullet \nabla \times \underline{u} - \underline{u} \bullet \nabla \times \underline{v}$$

## 3.2 *Series:*

**Standard Series:**

$$(b+x)^{\alpha} = b^{\alpha} + b^{\alpha-1}\alpha x + b^{\alpha-2}\,\frac{\alpha(\alpha-1)}{2!}x^2 + \ldots \text{ for } |x| < b$$

$$(1+x)^{\alpha} = 1 + \alpha x + \frac{\alpha(\alpha-1)}{2!}x^2 + \frac{\alpha(\alpha-1)(\alpha-2)}{3!}x^3 + \ldots$$

for arbitrary $\alpha, |x| < 1$

$$e^x = 1 + x + \frac{x^2}{2!} + \ldots + \frac{x^n}{n!} + \ldots \text{ for all } x$$

$$\cos x = 1 - \frac{x^2}{2!} + \frac{x^4}{4!} - \ldots + \frac{(-1)^n}{(2n)!}x^{2n} + \ldots \text{ for all } x$$

$$\sin x = x - \frac{x^3}{3!} + \frac{x^5}{5!} - \ldots + \frac{(-1)^n}{(2n+1)!}x^{2n+1} + \ldots \text{ for all } x$$

$$\tan x = x + \frac{x^3}{3} + \frac{2x^5}{15} + \frac{17x^7}{315} + \ldots \text{ for } |x| < \frac{\pi}{2}$$

$$\ln(1+x) = x - \frac{x^2}{2} + \frac{x^3}{3} - \ldots + \frac{(-1)^n}{(n+1)}x^{n+1} + \ldots \text{ for } -1 < x \le 1$$

**Taylor's Series**

$$f(a+h) = f(a) + hf'(a) + \frac{h^2}{2}f''(a) + \ldots$$

$$+ \frac{h^{n-1}}{(n-1)!}f^{(n-1)}(a) + \frac{h^n}{n!}f^{(n)}(c)$$

where $a < c < a+h$

**Maclaurin's Theorem:**

$$f(x) = f(0) + xf'(0) + \frac{x^2}{2!}f''(0) + \ldots$$

$$+ \frac{x^{n-1}}{(n-1)!}f^{(n-1)}(0) + \frac{x^n}{n!}f^{(n)}(\theta x)$$

where $0 < \theta < 1$

**Stirling's formula for $n!$:**

For large $n$:

$$n! \approx \sqrt{(2\pi)}\, n^{n+\frac{1}{2}} e^{-n}$$

or

$$\log_{10} n! \approx 0.39909 + (n + \tfrac{1}{2}) \log_{10} n - 0.43429n.$$

**Fourier Series:**

*General Formulae*:

If $f(x)$ is periodic, of period $2L$, $\quad f(x + 2L) = f(x)$

$$f(x) = \tfrac{1}{2} a_0 + \sum_{n=1}^{\infty} a_n \cos \frac{n\pi x}{L} + \sum_{n=1}^{\infty} b_n \sin \frac{n\pi x}{L}$$

where

$$a_n = \frac{1}{L} \int_{-L}^{L} f(x) \cos \frac{n\pi x}{L} \mathrm{d}x \qquad n = 0, 1, 2, \ldots$$

$$b_n = \frac{1}{L} \int_{-L}^{L} f(x) \sin \frac{n\pi x}{L} \mathrm{d}x \qquad n = 1, 2, 3, \ldots$$

If $f(x)$ is an even function of $x$ (i.e. $f(-x) = f(x)$), then

$$a_n = \frac{2}{L} \int_{0}^{L} f(x) \cos \frac{n\pi x}{L} \mathrm{d}x \qquad n = 0, 1, 2, \ldots$$

$$b_n = 0 \qquad n = 1, 2, 3, \ldots$$

If $f(x)$ is an odd function of $x$ (i.e. $f(-x) = -f(x)$), then

$$a_n = 0 \qquad n = 0, 1, 2, \ldots$$

$$b_n = \frac{2}{L} \int_{0}^{L} f(x) \sin \frac{n\pi x}{L} \mathrm{d}x \qquad n = 1, 2, 3, \ldots$$

*Special Waveforms, all of period 2L:*

**Square Wave, sine series**

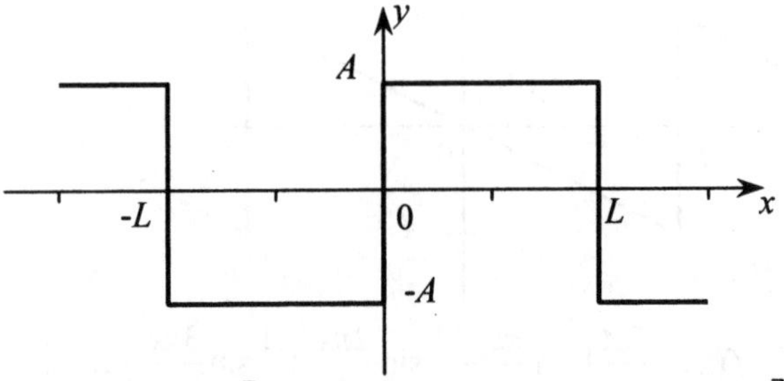


$$f(x) = \frac{4A}{\pi}\left[\sin\frac{\pi x}{L} + \frac{1}{3}\sin\frac{3\pi x}{L} + \frac{1}{5}\sin\frac{5\pi x}{L} + \ldots\right]$$

Mean square value $= A^2$

**Square Wave, cosine series**

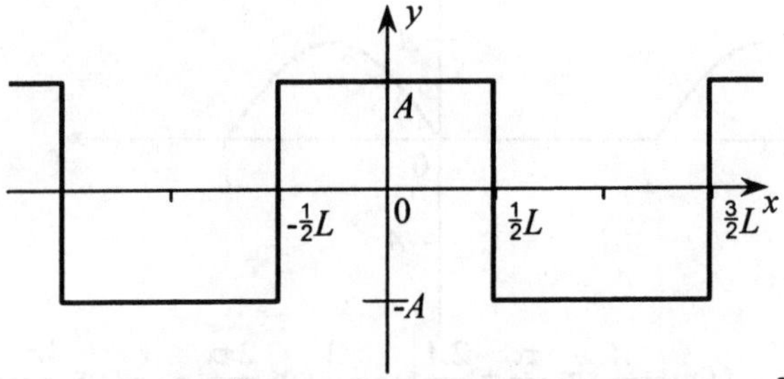


$$f(x) = \frac{4A}{\pi}\left[\cos\frac{\pi x}{L} - \frac{1}{3}\cos\frac{3\pi x}{L} + \frac{1}{5}\cos\frac{5\pi x}{L} - \ldots\right]$$

Mean square value $= A^2$

**Triangular Wave**

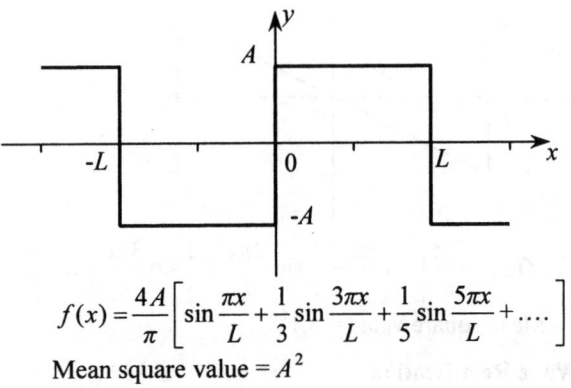

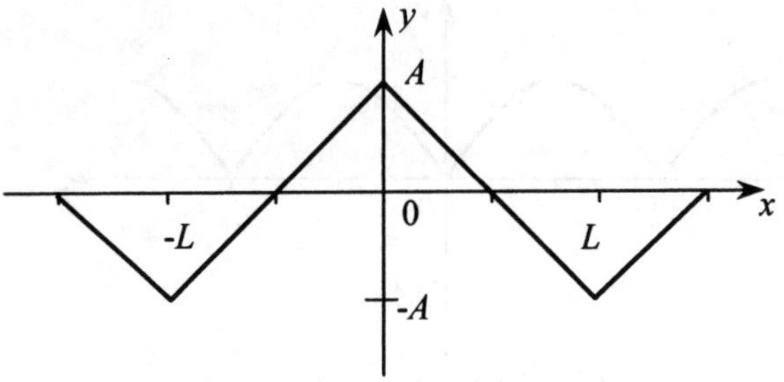


$$f(x) = \frac{8A}{\pi^2}\left[\cos\frac{\pi x}{L} + \frac{1}{3^2}\cos\frac{3\pi x}{L} + \frac{1}{5^2}\cos\frac{5\pi x}{L} + \ldots\right]$$

Mean square value $= \frac{1}{3}A^2$

**Saw-Tooth Wave**

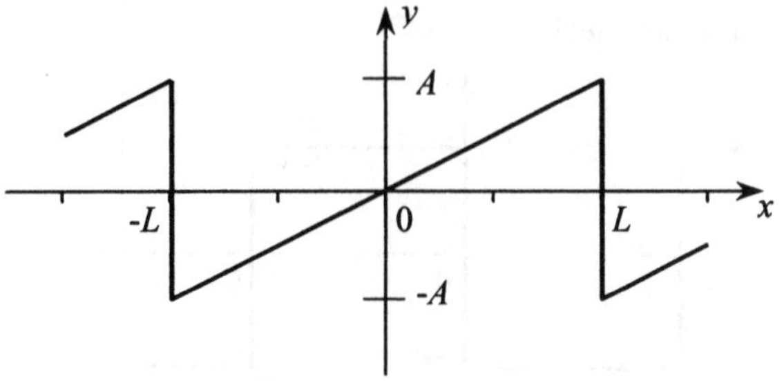


$$f(x)=\frac{2A}{\pi}\left[\sin\frac{\pi x}{L}-\frac{1}{2}\sin\frac{2\pi x}{L}+\frac{1}{3}\sin\frac{3\pi x}{L}-\ldots\right]$$

Mean square value = $\frac{1}{3}A^2$

**Half-Wave Rectification**

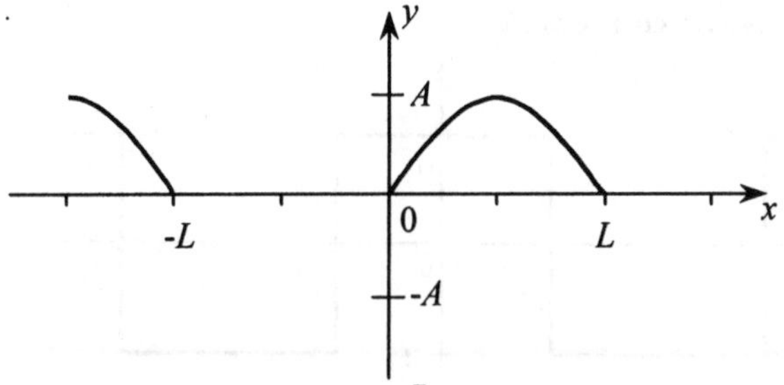


$$f(x)=\frac{A}{2}\sin\frac{\pi x}{L}+\frac{2A}{\pi}\left[\frac{1}{2}-\frac{1}{3}\cos\frac{2\pi x}{L}-\frac{1}{15}\cos\frac{4\pi x}{L}-\ldots\right]$$

Mean square value = $\frac{1}{4}A^2$, Average value $\frac{A}{\pi}$.

**Full-Wave Rectification**

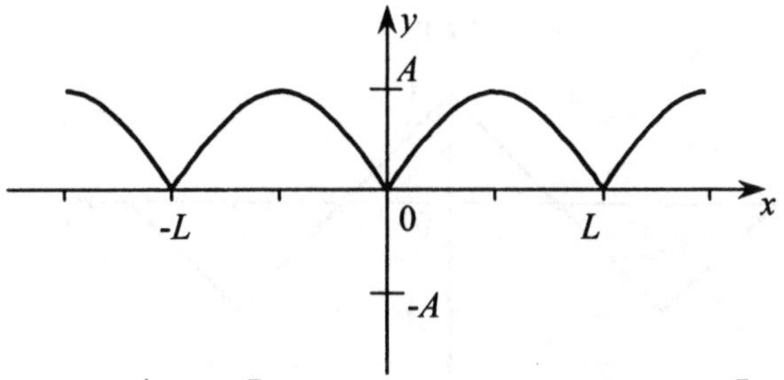


$$f(x)=\frac{4A}{\pi}\left[\frac{1}{2}-\frac{1}{3}\cos\frac{2\pi x}{L}-\frac{1}{15}\cos\frac{4\pi x}{L}-\ldots\right]$$

Mean square value = $\frac{1}{2}A^2$, Average value $\frac{2A}{\pi}$.

## 3.3 *Trigonometric, Hyperbolic and Algebraic Relations*

### Standard Trigonometrical Relationships:

$$\sin^2 A + \cos^2 A = 1$$
$$\sec^2 A = \tan^2 A + 1$$
$$2 \sin A \cos B = \sin(A - B) + \sin(A + B)$$
$$2 \cos A \cos B = \cos(A - B) + \cos(A + B)$$
$$2 \sin A \sin B = \cos(A - B) - \cos(A + B)$$
$$\sin A + \sin B = 2 \sin \tfrac{1}{2}(A + B) \cos \tfrac{1}{2}(A - B)$$
$$\sin A - \sin B = 2 \cos \tfrac{1}{2}(A + B) \sin \tfrac{1}{2}(A - B)$$
$$\cos A + \cos B = 2 \cos \tfrac{1}{2}(A + B) \cos \tfrac{1}{2}(A - B)$$
$$\cos A - \cos B = -2 \sin \tfrac{1}{2}(A + B) \sin \tfrac{1}{2}(A - B)$$
$$\sin(A \pm B) = \sin A \cos B \pm \cos A \sin B$$
$$\cos(A \pm B) = \cos A \cos B \mp \sin A \sin B$$
$$\tan(A \pm B) = \frac{\tan A \pm \tan B}{1 \mp \tan A \tan B}$$

**Positive Quadrants for Trigonometric Ratios:**

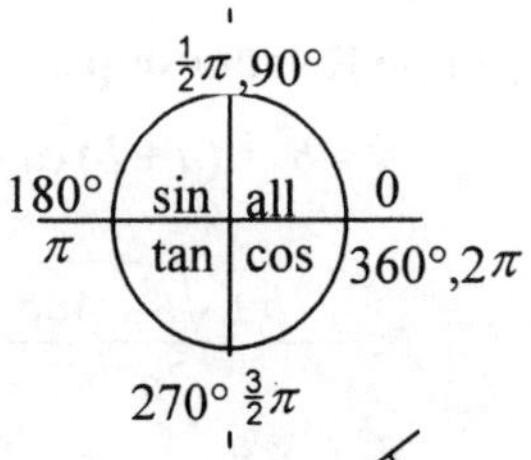


**Relationships for Plane Triangle:**

$$\frac{a}{\sin A} = \frac{b}{\sin B} = \frac{c}{\sin C}$$

$$\sin A = \frac{2}{bc}\sqrt{s(s - a)(s - b)(s - c)}$$

where $s = \frac{1}{2}(a + b + c)$

$$a^2 = b^2 + c^2 - 2bc \cos A$$
$$a^2 = b^2 + c^2 + 2bc \cos D$$

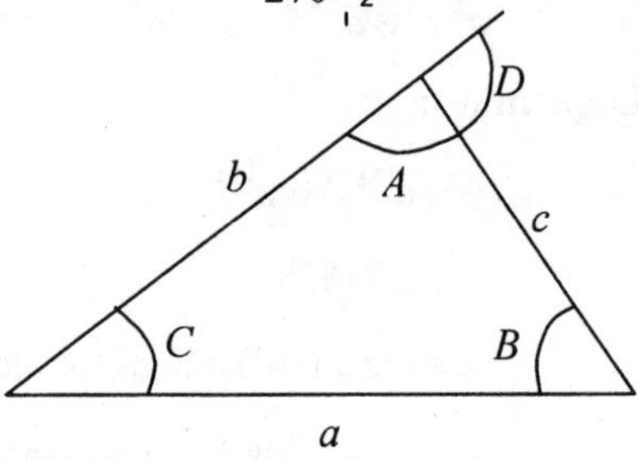


**Relationships for Spherical triangles**

$$\frac{\sin a}{\sin A} = \frac{\sin b}{\sin B} = \frac{\sin c}{\sin C}$$

$$\cos a = \cos b \cos c + \sin b \sin c \cos A$$

$$\cos A = -\cos B \cos C + \sin B \sin C \cos a$$

$$\sin \frac{A}{2} = \sqrt{\frac{\sin(s - b)\sin(s - c)}{\sin b \sin c}}$$

where $s = \frac{1}{2}(a + b + c)$

$$\sin \frac{a}{2} = \sqrt{-\frac{\cos S \cos(S - A)}{\sin B \sin C}}$$

where $S = \frac{1}{2}(A + B + C)$

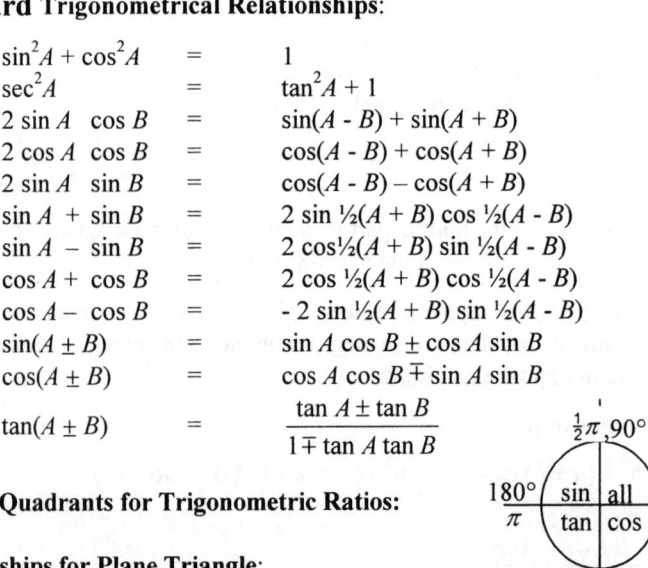

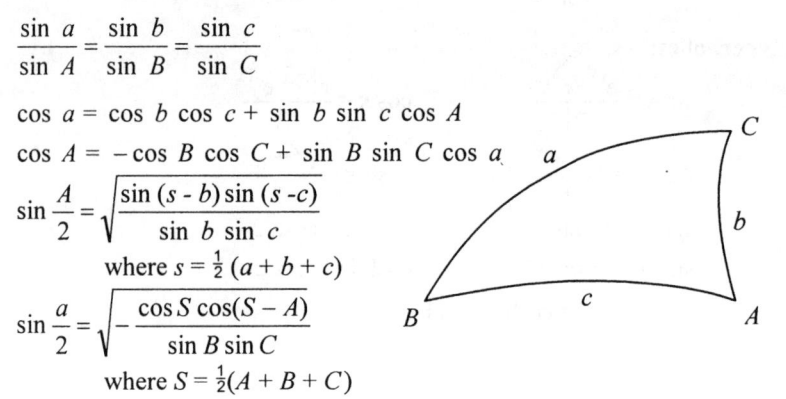

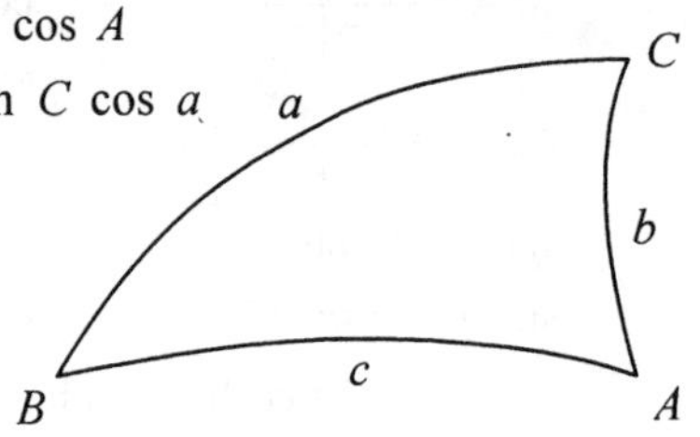

*Napier's Rules for right spherical triangles*:

Arrange the five parts about the right angle using the complement of the angle (co-$x$ = 90° - $x$) for the three parts opposite the right angle. E.g. If the right angle is at $A$ we have:

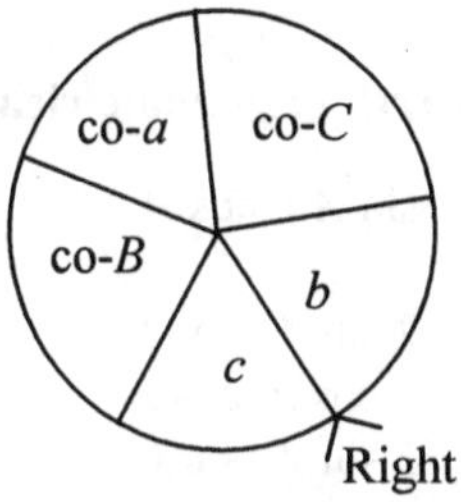


Then, the sine of the middle part is the product of the tangents of adjacent parts and is the product of the cosines of opposite parts.

N.B. A leg and its opposite angle are always in the same quadrant. If the hypotenuse is less than 90° the legs are in the same quadrant, otherwise they are in opposite quadrants.

**Algebraic Relationships:**

$$a^2 - b^2 = (a + b)(a - b) \qquad a^3 - b^3 = (a - b)(a^2 + ab + b^2)$$

$$x = \frac{-b \pm \sqrt{b^2 - 4ac}}{2a}$$

$$a^x a^y = a^{x+y} \qquad (a^x)^y = a^{xy}$$

**Logarithms:**

$$x \equiv e^{\log_e x} \equiv e^{\ln x}$$

$$e = 2.71828$$

$$x \equiv \log_{10}(10^x) \equiv \log_{10}(\text{antilog}_{10} x) \equiv 10^{\log_{10} x}$$

$$\log_e x = \frac{\log_{10} x}{\log_{10} e} = 2.30259 \log_{10} x$$

**Hyperbolics:**

$$\sin x = \frac{e^{ix} - e^{-ix}}{2i} \qquad \cos x = \frac{e^{ix} + e^{-ix}}{2}$$

$$\sinh x = \frac{e^x - e^{-x}}{2} \qquad \cosh x = \frac{e^x + e^{-x}}{2}$$

$$\sin iz = i \sinh z \qquad \cos iz = \cosh z$$

$$\sinh iz = i \sin z \qquad \cosh iz = \cos z$$

$$e^z = \cosh z + \sinh z$$

**Equations of Curves:**

| Circle | Ellipse | Hyperbola | Parabola |
|---|---|---|---|
| $x^2 + y^2 = a^2$ | $\frac{x^2}{a^2} + \frac{y^2}{b^2} = 1$ | $\frac{x^2}{a^2} - \frac{y^2}{b^2} = 1$ | $y^2 = ax$ |

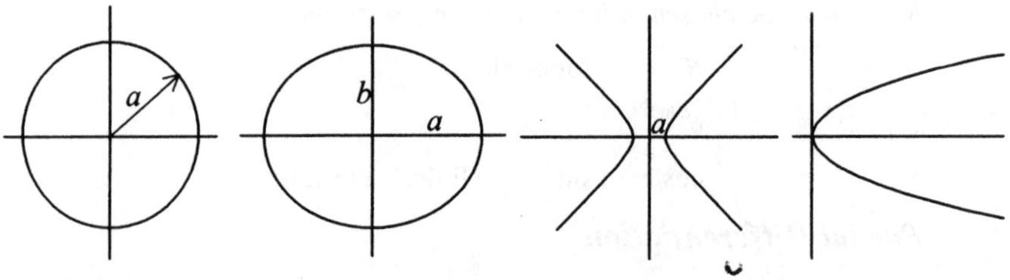


**Standard Curves of Trigonometrical Functions**

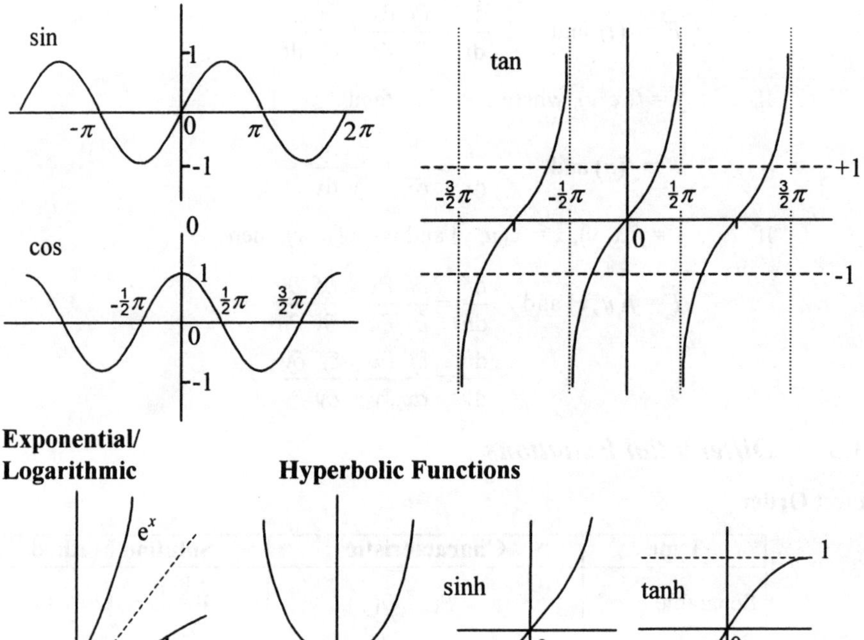


**Exponential/ Logarithmic**

**Hyperbolic Functions**

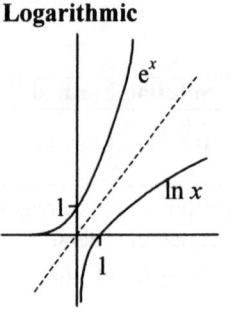

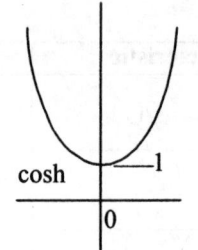

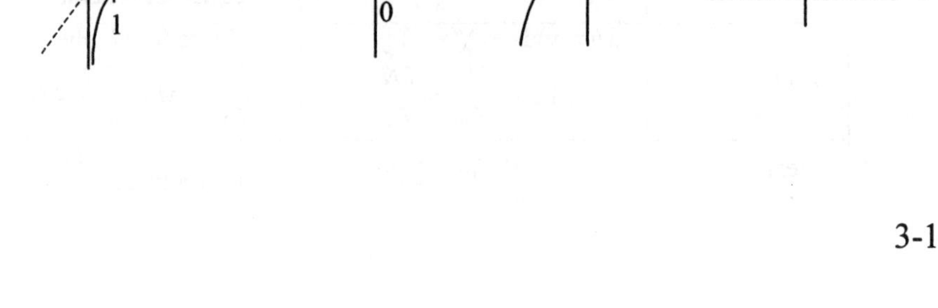

## 3.4 *Complex Numbers*

$$z = x + \mathrm{i}\,y = r(\cos\theta + \mathrm{i}\sin\theta)$$
$$= r\,\mathrm{e}^{\mathrm{i}(\theta + 2n\pi)} \qquad n = 0, \pm1, \pm2, \ldots$$
$$r = \sqrt{x^2 + y^2} \qquad \theta = \tan^{-1}\left(\frac{y}{x}\right)$$

*N.B. $\theta$ must be chosen to lie in the appropriate quadrant.*

$$|z - z_0| = R \qquad \text{for circle}$$
$$z^c = \mathrm{e}^{c \ln z}$$
$$\mathrm{e}^{\mathrm{i}z} = \cos z + \mathrm{i}\sin z \qquad \text{(Euler's Formula)}$$

## 3.5 *Partial Differentiation*

If $f = f(x, y)$, where $x = x(t)$ and $y = y(t)$, then

$$F = f(t) \text{ and } \quad \frac{\mathrm{d}f}{\mathrm{d}t} = \frac{\partial f}{\partial x}\frac{\mathrm{d}x}{\mathrm{d}t} + \frac{\partial f}{\partial y}\frac{\mathrm{d}y}{\mathrm{d}t}$$

If $f = f(x, y)$, where $y = y(x)$, then

$$F = f(x) \text{ and } \quad \frac{\mathrm{d}f}{\mathrm{d}x} = \frac{\partial f}{\partial x} + \frac{\partial f}{\partial y}\frac{\mathrm{d}y}{\mathrm{d}x}$$

If $f = f(x, y)$, $x = x(u, v)$ and $y = y(u, v)$, then

$$F = f(u, v) \text{ and } \quad \frac{\mathrm{d}f}{\mathrm{d}u} = \frac{\partial f}{\partial x}\frac{\partial x}{\partial u} + \frac{\partial f}{\partial y}\frac{\partial y}{\partial u}$$
$$\frac{\mathrm{d}f}{\mathrm{d}v} = \frac{\partial f}{\partial x}\frac{\partial x}{\partial v} + \frac{\partial f}{\partial y}\frac{\partial y}{\partial v}$$

## 3.6 *Differential Equations*

**First Order**

| Type | Characteristic | Solution Method |
|---|---|---|
| Separable | $y' = P(x)\,Q(y)$ | $\int \frac{1}{Q}\mathrm{d}y = \int P\mathrm{d}x + c$ |
| Homogeneous | $y' = f\left(\frac{y}{x}\right)$ | substitution $y = ux$ makes equation separable |
| Exact | $M(x, y)\mathrm{d}x + N(x, y)\mathrm{d}y = 0$ where $\frac{\partial M}{\partial y} = \frac{\partial N}{\partial x}$ | Solve for $G$ where $\frac{\partial G}{\partial x} = M$, $\frac{\partial G}{\partial y} = N$ |
| Linear | $y' + P(x)y = Q(x)$ | multiply by $\mathrm{e}^{\int P\mathrm{d}x}$ |

**Second Order**, linear with constant coefficients

*General Form:*

$$\frac{d^2y}{dx^2} + a\frac{dy}{dx} + by = f(x)$$

**General Solution** = **Complementary Function** + **Particular Integral**

<u>*or*</u>

Dynamic Response (GS) = Steady State (CF) + Transient Response (PI)

| **G.S. = C.F. + P.I.** | | |
|---|---|---|
| **C.F. Transient** | Auxiliary equation: $m^2 + am + b = (m - m_1)(m - m_2) = 0$ | |
| | **Roots:** $m_1$ **and** $m_2$ | **C.F.** |
| | $m_1, m_2$ real, $m_1 \neq m_2$ | $A_1 e^{m_1 x} + A_2 e^{m_2 x}$ |
| | $m_1 = m_2$ | $(A + Bx)e^{m_1 x}$ |
| | $m_1, m_2$ complex conjugate $(\alpha \pm i\beta)$ | $Ce^{\alpha x} \cos(\beta x - \varepsilon)$ |
| **P.I. Steady State** | **Forcing function:** $f(x)$ | **P.I.** |
| | $f(x)$ is polynomial | Polynomial of same degree |
| | $f(x) = \lambda e^{\alpha x}$ | |
| | $\alpha \neq m_1$ or $m_2$ | $Ae^{\alpha x}$ |
| | $\alpha = m_1 \neq m_2$ | $Axe^{\alpha x}$ |
| | $\alpha = m_1 = m_2$ | $Ax^2e^{\alpha x}$ |
| | $f(x) = F \cos \alpha x$ | Re $(Ae^{i\alpha x})$ |
| | $f(x) = F \sin \alpha x$ | Im $(Ae^{i\alpha x})$ |

Re( ) = real part,
Im( ) = imaginary part

*Transient behaviour: free vibration*

$$m\ddot{x} + \alpha\dot{x} + kx = 0 \qquad \text{or} \qquad \ddot{x} + 2\xi\omega_0\dot{x} + \omega_0^2 x = 0$$

$$\text{where } \xi = \frac{\alpha}{2\sqrt{mk}} = \frac{\alpha}{2m\omega_0}, \quad \omega_0 = \sqrt{\frac{k}{m}}$$

$\xi < 1$ **(underdamping)**

$x = a\,\mathrm{e}^{-\xi\omega_0 t}\cos(\omega t - \theta)$

where $\omega = \omega_0\sqrt{1-\xi^2}$

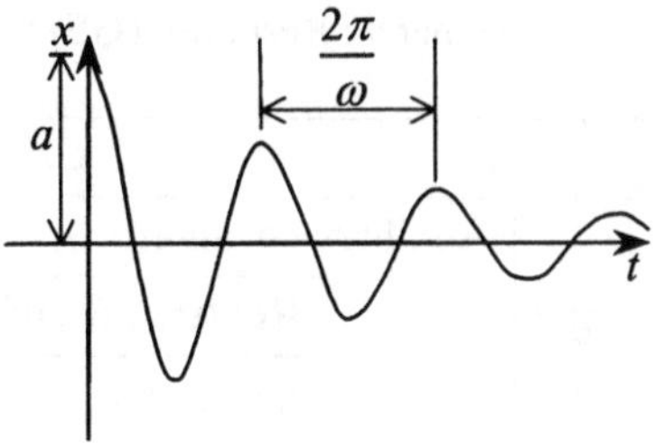


Ratio of successive peaks:

$$R = \mathrm{e}^{\frac{2\pi\xi}{\sqrt{1-\xi^2}}}$$

Logarithmic decrement:

$$d = \ln R' = \frac{2\pi\xi}{\sqrt{1-\zeta^2}} \approx 2\pi\xi \text{ for small } \xi.$$

$\xi > 1$ **(overdamping)**

$x = A_1\,\mathrm{e}^{-q_1 t} + A_2\,\mathrm{e}^{-q_2 t}$

where $q_1, q_2 = \omega_0(\xi \pm \sqrt{\xi^2 - 1})$
and $(q_1 > q_2)$

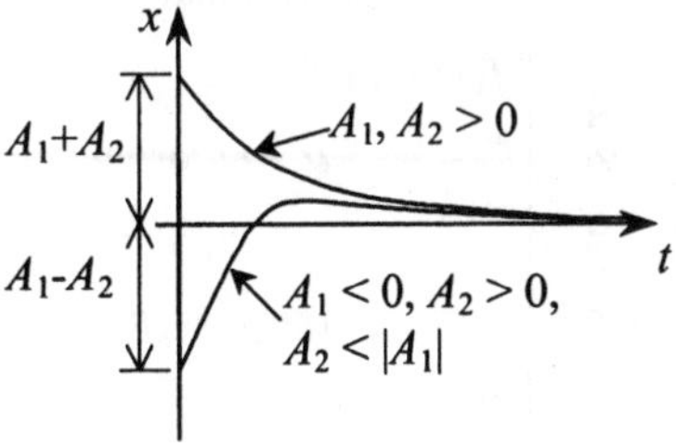


$\xi = 1$ **(critical damping)**

$x = (A + Bt)\mathrm{e}^{-\xi\omega_0 t}$

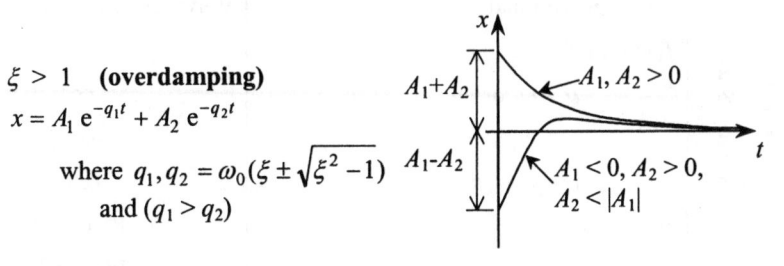

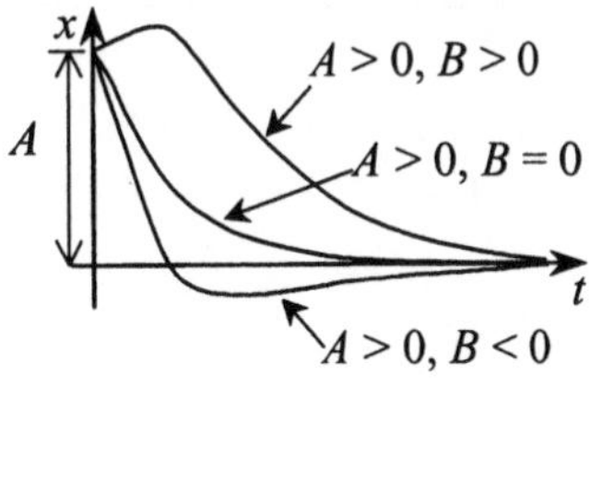

**Constant amplitude forcing**

$$m\ddot{x} + \alpha\dot{x} + kx = F\cos pt \qquad \text{or} \qquad \ddot{x} + 2\xi\omega_0\dot{x} + \omega_0^2 x = \frac{F}{m}\cos pt$$

$$\text{where } \xi = \frac{\alpha}{2\sqrt{mk}} = \frac{\alpha}{2m\omega_0}, \quad \omega_0 = \sqrt{\frac{k}{m}}$$

Particular integral is:

$$x = \frac{AF}{m\omega_0^2}\cos(pt - \phi) \qquad \text{or} \qquad x = Ax_1\cos(pt - \phi)$$

$$\text{where} \quad x_1 = \frac{F}{k}, \quad \tan\phi = \frac{2\xi\frac{p}{\omega_0}}{1-\left(\frac{p}{\omega_0}\right)^2},$$

$$\text{and} \qquad A = \left|\frac{x}{x_1}\right| = \frac{1}{\left[\left(1-\left(\frac{p}{\omega_0}\right)^2\right)^2 + \left(2\xi\frac{p}{\omega_0}\right)^2\right]^{\frac{1}{2}}}$$

At resonance, $|x|$ is a maximum, and

$$\frac{x}{x_1} = \frac{1}{2\xi\sqrt{1-\xi^2}}, \quad p = \omega_0\sqrt{1-2\xi^2}, \quad \tan\phi = \frac{\sqrt{1-2\xi^2}}{\xi}$$

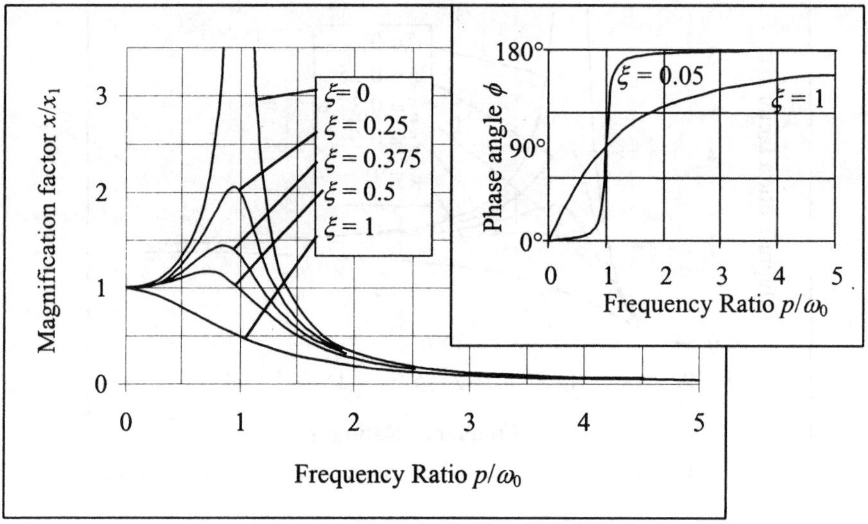

**$p^2$ Forcing (out of balance mass etc.)**

$$m\ddot{x} + \alpha\dot{x} + kx = mx_1 p^2 \cos pt \quad \text{or} \quad \ddot{x} + 2\xi\omega_0\dot{x} + \omega_0^2 x = x_1 p^2 \cos pt$$

$$\text{where } \xi = \frac{\alpha}{2\sqrt{mk}} = \frac{\alpha}{2m\omega_0}, \quad \omega_0 = \sqrt{\frac{k}{m}}$$

Particular integral is:

$$x = Ax_1 \cos(pt - \phi)$$

$$\text{where} \quad \tan\phi = \frac{2\xi \dfrac{p}{\omega_0}}{1 - \left(\dfrac{p}{\omega_0}\right)^2}$$

$$\text{and} \quad A = \left|\frac{x}{x_1}\right| = \frac{\left(\dfrac{p}{\omega_0}\right)^2}{\left[\left(1 - \left(\dfrac{p}{\omega_0}\right)^2\right)^2 + \left(2\xi\dfrac{p}{\omega_0}\right)^2\right]^{\frac{1}{2}}}$$

At resonance, $|x|$ is a maximum, and

$$\frac{x}{x_1} = \frac{1}{2\xi\sqrt{1-\xi^2}}, \quad p = \frac{\omega_0}{\sqrt{1-2\xi^2}}, \quad \tan\phi = \frac{\sqrt{1-2\xi^2}}{\xi}$$

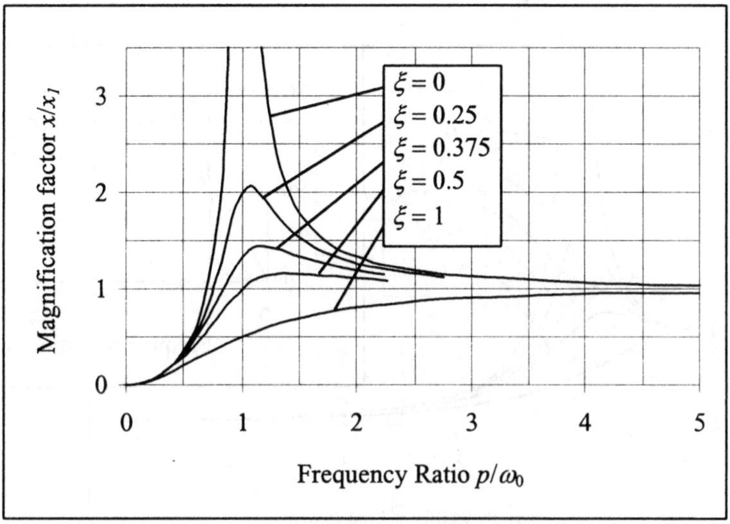

## 3.7 *Rules for Differentiation and Integration*

| | |
|---|---|
| $\frac{\mathrm{d}}{\mathrm{d}x}(uv) = u\frac{\mathrm{d}v}{\mathrm{d}x} + v\frac{\mathrm{d}u}{\mathrm{d}x}$ | $\frac{\mathrm{d}}{\mathrm{d}x}(uvw) = uv\frac{\mathrm{d}w}{\mathrm{d}x} + uw\frac{\mathrm{d}v}{\mathrm{d}x} + vw\frac{\mathrm{d}u}{\mathrm{d}x}$ |
| $\frac{\mathrm{d}}{\mathrm{d}x}\left(\frac{u}{v}\right) = \frac{1}{v^2}\left(v\frac{\mathrm{d}u}{\mathrm{d}x} - u\frac{\mathrm{d}v}{\mathrm{d}x}\right)$ | $\int uv\,\mathrm{d}x = uw - \int\frac{\mathrm{d}u}{\mathrm{d}x}w\,\mathrm{d}x,\quad \text{where } w = \int v\,\mathrm{d}x$ |

## 3.8 *Standard Differentials and Integrals*

(Note: all the indefinite integrals below also require a constant of integration)

$$\frac{\mathrm{d}}{\mathrm{d}x}x^n = nx^{n-1} \qquad \int x^n\mathrm{d}x = \frac{x^{n+1}}{n+1},\quad n \neq 1$$

$$\frac{\mathrm{d}}{\mathrm{d}x}\ln|x| = \frac{1}{x} \qquad \int\frac{\mathrm{d}x}{x} = \ln|x|$$

$$\frac{\mathrm{d}}{\mathrm{d}x}\mathrm{e}^{ax} = a\mathrm{e}^{ax} \qquad \int \mathrm{e}^{ax}\mathrm{d}x = \frac{\mathrm{e}^{ax}}{a},\quad a \neq 0$$

$$\frac{\mathrm{d}}{\mathrm{d}x}a^x = a^x\ln a \qquad \int a^x\mathrm{d}x = \frac{a^x}{\ln a},\quad a > 0, a \neq 1$$

$$\frac{\mathrm{d}}{\mathrm{d}x}x^x = x^x(1+\ln x) \qquad \int \ln x\,\mathrm{d}x = x(\ln x - 1)$$

$$\frac{\mathrm{d}}{\mathrm{d}x}\sin x = \cos x \qquad \int \sin x\,\mathrm{d}x = -\cos x$$

$$\frac{\mathrm{d}}{\mathrm{d}x}\cos x = -\sin x \qquad \int \cos x\,\mathrm{d}x = \sin x$$

$$\frac{\mathrm{d}}{\mathrm{d}x}\tan x = \sec^2 x \qquad \int \tan x\,\mathrm{d}x = -\ln|\cos x|$$

$$\frac{\mathrm{d}}{\mathrm{d}x}\cot x = -\mathrm{cosec}^2 x \qquad \int \cot x\,\mathrm{d}x = \ln|\sin x|$$

$$\frac{\mathrm{d}}{\mathrm{d}x}\sin^{-1}x = \frac{1}{\sqrt{1-x^2}} \qquad \int \sec^2 x\,\mathrm{d}x = \tan x$$

$$\frac{\mathrm{d}}{\mathrm{d}x}\cos^{-1}x = \frac{-1}{\sqrt{1-x^2}} \qquad \int \mathrm{cosec}^2 x\,\mathrm{d}x = -\cot x$$

$$\frac{\mathrm{d}}{\mathrm{d}x}\tan^{-1}x = \frac{1}{1+x^2} \qquad \int\frac{\mathrm{d}x}{\sqrt{1-x^2}} = \sin^{-1}x,\quad |x| < 1$$

$$\frac{\mathrm{d}}{\mathrm{d}x}\cot^{-1}x = -\frac{1}{1+x^2} \qquad \int\frac{\mathrm{d}x}{1+x^2} = \tan^{-1}x$$

$$\frac{\mathrm{d}}{\mathrm{d}x}\sinh x = \cosh x \qquad \int \cosh x\,\mathrm{d}x = \sinh x$$

$$\frac{d}{dx}\cosh x = \sinh x \qquad \int \sinh x \, dx = \cosh x$$

$$\frac{d}{dx}\tanh x = \operatorname{sech}^2 x \qquad \int \operatorname{sech}^2 x \, dx = \tanh x$$

$$\frac{d}{dx}\coth x = -\operatorname{cosech}^2 x \qquad \int \operatorname{cosech}^2 x \, dx = -\coth x$$

$$\frac{d}{dx}\sinh^{-1} x = \frac{1}{\sqrt{x^2+1}} \qquad \int \frac{dx}{\sqrt{x^2+1}} = \sinh^{-1} x = \ln\left|x+\sqrt{x^2+1}\right|$$

$$\frac{d}{dx}\cosh^{-1} x = \frac{1}{\sqrt{x^2-1}} \qquad \int \frac{dx}{\sqrt{x^2-1}} = \cosh^{-1} x = \ln\left|x+\sqrt{x^2-1}\right| \quad x \ge 1$$

$$\frac{d}{dx}\tanh^{-1} x = \frac{1}{1-x^2}, \quad x^2<1 \qquad \int \frac{dx}{1-x^2} = \tanh^{-1} x = \tfrac{1}{2}\ln\left|\frac{1+x}{1-x}\right| \quad x^2<1$$

$$\frac{d}{dx}\coth^{-1} x = \frac{1}{1-x^2}, \quad x^2>1 \qquad \int \frac{dx}{x^2-1} = \coth^{-1} x = \tfrac{1}{2}\ln\left|\frac{x-1}{x+1}\right| \quad x^2>1$$

**Some Definite Integrals** (*m* and *n* are integers in the following)

$$\int_0^{\frac{\pi}{2}} \sin^n x \, dx = \int_0^{\frac{\pi}{2}} \cos^n x \, dx = \begin{cases} \dfrac{n-1}{n}\times\dfrac{n-3}{n-2}\cdots\dfrac{3}{4}\times\dfrac{1}{2}\times\dfrac{\pi}{2}, & n \text{ even} \\ \dfrac{n-1}{n}\times\dfrac{n-3}{n-2}\cdots\dfrac{4}{5}\times\dfrac{2}{3}\times 1, & n \text{ odd} \end{cases}$$

$$I_{m,n} = \int_0^{\frac{\pi}{2}} \sin^m x \cos^n x \, dx = \left(\frac{m-1}{m+n}\right) I_{m-2,n} = \left(\frac{n-1}{m+n}\right) I_{m,n-2}, \quad m \ne -n$$

$$\int_0^{\pi} \sin mx \sin nx \, dx = \int_0^{\pi} \cos mx \cos nx \, dx = 0, \quad m \ne n$$

$$\int_0^{\infty} e^{-ax}\sin bx \, dx = \frac{b}{a^2+b^2} \quad (a>0) \qquad \int_0^{\infty} e^{-ax}\cos bx \, dx = \frac{a}{a^2+b^2} \quad (a>0)$$

$$\int_0^{\pi} \frac{\cos nx}{\cos x - \cos\phi} dx = \frac{\pi \sin n\phi}{\sin\phi} \qquad \int_0^{\pi} \sin nx \cos nx \, dx = 0$$

$$\int_0^{\pi} \frac{\sin x \cos x}{(1+a\cos x)^3} dx = \frac{-2a}{(1-a^2)^2} \qquad \int_0^{2\pi} \frac{dx}{(1+a\cos x)} = \frac{2\pi}{\sqrt{(1-a^2)}}$$

$$\int_0^{\infty} e^{-x^2} dx = \frac{\sqrt{\pi}}{2} \qquad \int_0^{\pi} \frac{\sin^2 x}{(1+a\cos x)^3} dx = \frac{\pi}{2(1-a^2)^{\frac{3}{2}}}$$

## 3.9 ***The Error Function - Selected Values:***

Definition: $$\operatorname{erf} z = \frac{2}{\sqrt{\pi}} \int_0^z e^{-u^2} \, du$$

| $z$ | erf $z$ | $z$ | erf $z$ | $z$ | erf $z$ | $z$ | erf $z$ |
|---|---|---|---|---|---|---|---|
| 0.00 | 0.0000 | 0.62 | 0.6194 | 1.24 | 0.9205 | 1.86 | 0.9915 |
| 0.02 | 0.0226 | 0.64 | 0.6346 | 1.26 | 0.9252 | 1.88 | 0.9922 |
| 0.04 | 0.0451 | 0.66 | 0.6494 | 1.28 | 0.9297 | 1.90 | 0.9928 |
| 0.06 | 0.0676 | 0.68 | 0.6638 | 1.30 | 0.9340 | 1.92 | 0.9934 |
| 0.08 | 0.0901 | 0.70 | 0.6778 | 1.32 | 0.9381 | 1.94 | 0.9939 |
| 0.10 | 0.1125 | 0.72 | 0.6914 | 1.34 | 0.9419 | 1.96 | 0.9944 |
| 0.12 | 0.1348 | 0.74 | 0.7047 | 1.36 | 0.9456 | 1.98 | 0.9949 |
| 0.14 | 0.1569 | 0.76 | 0.7175 | 1.38 | 0.9490 | 2.00 | 0.9953 |
| 0.16 | 0.1790 | 0.78 | 0.7300 | 1.40 | 0.9523 | 2.05 | 0.9963 |
| 0.18 | 0.2009 | 0.80 | 0.7421 | 1.42 | 0.9554 | 2.10 | 0.9970 |
| 0.20 | 0.2227 | 0.82 | 0.7538 | 1.44 | 0.9583 | 2.15 | 0.9976 |
| 0.22 | 0.2443 | 0.84 | 0.7651 | 1.46 | 0.9611 | 2.20 | 0.9981 |
| 0.24 | 0.2657 | 0.86 | 0.7761 | 1.48 | 0.9637 | 2.25 | 0.9985 |
| 0.26 | 0.2869 | 0.88 | 0.7867 | 1.50 | 0.9661 | 2.30 | 0.9989 |
| 0.28 | 0.3079 | 0.90 | 0.7969 | 1.52 | 0.9684 | 2.35 | 0.9991 |
| 0.30 | 0.3286 | 0.92 | 0.8068 | 1.54 | 0.9706 | 2.40 | 0.9993 |
| 0.32 | 0.3491 | 0.94 | 0.8163 | 1.56 | 0.9726 | 2.45 | 0.9995 |
| 0.34 | 0.3694 | 0.96 | 0.8254 | 1.58 | 0.9745 | 2.50 | 0.9996 |
| 0.36 | 0.3893 | 0.98 | 0.8342 | 1.60 | 0.9763 | 2.55 | 0.9997 |
| 0.38 | 0.4090 | 1.00 | 0.8427 | 1.62 | 0.9780 | 2.60 | 0.9998 |
| 0.40 | 0.4284 | 1.02 | 0.8508 | 1.64 | 0.9796 | 2.65 | 0.9998 |
| 0.42 | 0.4475 | 1.04 | 0.8586 | 1.66 | 0.9811 | 2.70 | 0.9999 |
| 0.44 | 0.4662 | 1.06 | 0.8661 | 1.68 | 0.9825 | 2.75 | 0.9999 |
| 0.46 | 0.4847 | 1.08 | 0.8733 | 1.70 | 0.9838 | 2.80 | 0.9999 |
| 0.48 | 0.5027 | 1.10 | 0.8802 | 1.72 | 0.9850 | 2.85 | 0.9999 |
| 0.50 | 0.5205 | 1.12 | 0.8868 | 1.74 | 0.9861 | 2.90 | 1.0000 |
| 0.52 | 0.5379 | 1.14 | 0.8931 | 1.76 | 0.9872 | 2.95 | 1.0000 |
| 0.54 | 0.5549 | 1.16 | 0.8991 | 1.78 | 0.9882 | 3.00 | 1.0000 |
| 0.56 | 0.5716 | 1.18 | 0.9048 | 1.80 | 0.9891 | 4.00 | 1.0000 |
| 0.58 | 0.5879 | 1.20 | 0.9103 | 1.82 | 0.9899 | | |
| 0.60 | 0.6039 | 1.22 | 0.9155 | 1.84 | 0.9907 | | |

## 3.10 *Laplace Transforms*

Definition: $F(s) = \mathrm{L}[f(t)] = \int_0^\infty f(t)\,\mathrm{e}^{-st}\,\mathrm{d}t$

**Theorems:**

| | | |
|---|---|---|
| Linearity | $\mathrm{L}[af(t) + bg(t)]$ | $= aF(s) + bG(s)$ |
| Final Value | $\lim_{t\to\infty} f(t)$ | $= \lim_{s\to 0} sF(s)$ |
| Initial Value | $\lim_{t\to 0} f(t)$ | $= \lim_{s\to\infty} sF(s)$ |
| Differentiation | $\mathrm{L}\left[\frac{\mathrm{d}f(t)}{\mathrm{d}t}\right]$ | $= sF(s) - f(0)$ |
| | $\mathrm{L}\left[\frac{\mathrm{d}^2 f(t)}{\mathrm{d}t^2}\right]$ | $= s^2F(s) - sf(0) - \frac{\mathrm{d}f(0)}{\mathrm{d}t}$ |
| Integration | $\mathrm{L}\left[\int f(t)\mathrm{d}t\right]$ | $= \frac{F(s)}{s} + \frac{f^{-1}(0)}{s}$ |
| First Shifting | $\mathrm{L}[\mathrm{e}^{at}f(t)]$ | $= F(s - a)$ |
| Second Shifting | $\mathrm{L}[f(t-a)]$ | $= \mathrm{e}^{-as}F(s) \quad (t > a)$ |
| Convolution: $\mathrm{L}[f*g] \equiv$ | $\mathrm{L}\left[\int_0^t f(u)g(t-u)\mathrm{d}u\right]$ | $= F(s)G(s)$ |
| Partial Differentiation | $\mathrm{L}\left[\frac{\partial f(t,\alpha)}{\partial \alpha}\right]$ | $= \frac{\partial}{\partial \alpha}F(s,\alpha)$ |
| Time Multiplication | $\mathrm{L}[tf(t)]$ | $= -\frac{\mathrm{d}F(s)}{\mathrm{d}s}$ |

**Transform Pairs**

| Function | Laplace Transform |
|---|---|
| $1$ | $\frac{1}{s}$ |
| $H(t-T) = 0 \quad t < T$<br>$= 1 \quad t \geq T$ | $\frac{1}{s}\mathrm{e}^{-sT}$ |
| $t^n$ | $\frac{n!}{s^{n+1}}$ |

| Function | Laplace Transform |
|---|---|
| $e^{-at}$ | $\dfrac{1}{s+a}$ |
| $\sin \omega t$ | $\dfrac{\omega}{s^2+\omega^2}$ |
| $\cos \omega t$ | $\dfrac{s}{s^2+\omega^2}$ |
| $1-e^{-t/T}$ | $\dfrac{1}{s(1+Ts)}$ |
| $\dfrac{\omega_n}{\sqrt{1-\xi^2}} e^{-\xi\omega_n t} \sin\left(\omega_n \sqrt{1-\xi^2 t}\right)$ | $\dfrac{1}{\left(1+2\dfrac{\xi s}{\omega_n}+\dfrac{s^2}{\omega_n^2}\right)}$ |
| $1-\dfrac{1}{\sqrt{1-\xi^2}} e^{-\xi\omega_n t} \sin\left(\omega_n \sqrt{1-\xi^2} t+\cos^{-1}\xi\right)$ | $\dfrac{1}{s\left(1+2\dfrac{\xi s}{\omega_n}+\dfrac{s^2}{\omega_n^2}\right)}$ |

## 3.11 *Numerical Analysis*

**Approximate solution of an Algebraic Equation** $f(x) = 0$

*Newton's Method*

$$x_1 = x_0 - \frac{f(x_0)}{f'(x_0)}$$

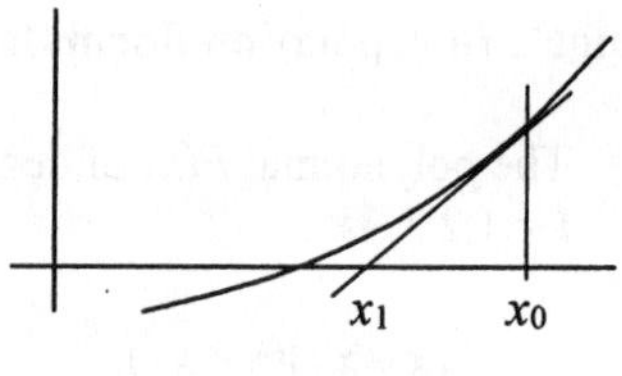


*Secant Method*

$$x_1 = \frac{-x_0 f(x_{-1}) + x_{-1} f(x_0)}{f(x_0) - f(x_{-1})}$$

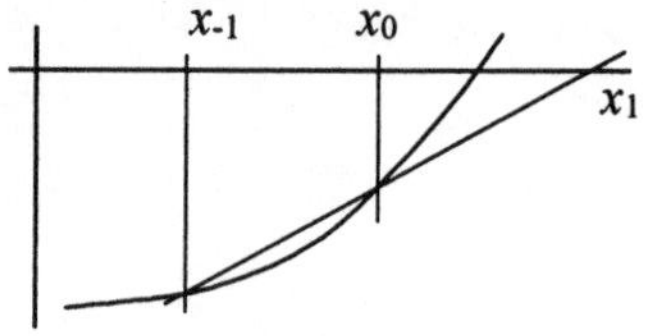

**Least Squares curve fitting of a straight line**

If $y_i$ ($i = 1, 2, \ldots n$) are experimental values of $y$ at chosen exact values of $x_i$, then the line of best fit passes through the centroid of the points:

$$\bar{x} = \frac{1}{n}\sum_{i=1}^{n} x_i \qquad \bar{y} = \frac{1}{n}\sum_{i=1}^{n} y_i$$

and is given by $y = mx + c$, where

$$m = \frac{\sum (x_i - \bar{x})(y_i - \bar{y})}{\sum (x_i - \bar{x})^2} = \frac{\sum x_i y_i - n\bar{x}\bar{y}}{\sum x_i^2 - n\bar{x}^2}$$

$$c = \bar{y} - m\bar{x}$$

**Finite Difference Formulae**

$$\Delta f(x) = f(x + h) - f(x)$$

$$f'(x) = \frac{f(x+h) - f(x-h)}{2h} + O(h^2)$$

$$f''(x) = \frac{f(x+h) - 2f(x) + f(x-h)}{h^2} + O(h^2)$$

$$f'''(x) = \frac{f(x+2h) - 2f(x+h) + 2f(x-h) - f(x-2h)}{2h^3}$$

**Lagrange's Interpolation Formula** for unequal intervals

The polynomial $P(x)$ of degree 2 passing through three points $(x_i, y_i)$, $i = 1,2,3$, is

$$P(x) = \frac{(x-x_2)(x-x_3)}{(x_1-x_2)(x_1-x_3)} y_1 + \frac{(x-x_1)(x-x_3)}{(x_2-x_1)(x_2-x_3)} y_2 + \frac{(x-x_1)(x-x_2)}{(x_3-x_1)(x_3-x_2)} y_3$$

**Formulae for Numerical Integration**

Divide the range into equal intervals of width $h$ such that

$x_n = x_0 + nh$

$y_n = y_n(x_n)$

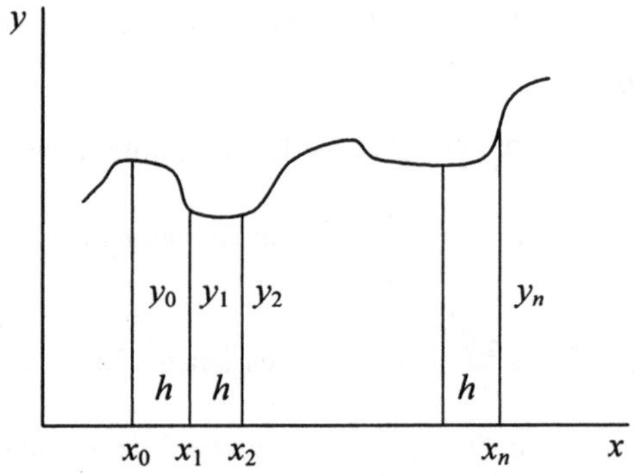


*Trapezoidal Rule:*

1 strip: $$\int_{x_0}^{x_1} y(x)dx = \frac{h}{2}[y_0 + y_1] + \varepsilon, \quad \text{where } \varepsilon \approx \frac{-h^3}{12} y_0''$$

$n$ strips: $$\int_{x_0}^{x_n} y(x)dx \approx \frac{h}{2}[y_0 + 2y_1 + 2y_2 + \ldots + 2y_{n-1} + y_n]$$

*Simpson's Rule*:

2 strips: $$\int_{x_0}^{x_2} y(x)dx = \frac{h}{3}[y_0 + 4y_1 + y_2] + \varepsilon, \quad \text{where } \varepsilon \approx \frac{-h^5}{90} y_1^{(4)}$$

$n$ strips ($n$ even):

$$\int_{x_0}^{x_n} y(x)dx \approx \frac{h}{3}[y_0 + y_n + 4(y_1 + y_3 + \ldots + y_{n-1}) + 2(y_2 + y_4 + \ldots + y_{n-2})]$$

= ($h$/3) × ("first + last + four×odds + two×evens")

*Runge-Kutta Method*:

Second Order: $$y_{n+1} = y_n + \frac{h}{2}\{f(x_n, y_n) + f(x_n + h, y_n + k_1)\}$$

Fourth Order: $$y_{n+1} = y_n + \frac{1}{6}(k_1 + 2k_2 + 2k_3 + k_4)$$

$$k_1 = hf(x_n, y_n)$$

$$k_2 = hf\left(x_n + \frac{h}{2}, y_n + \frac{k_1}{2}\right)$$

$$k_3 = hf\left(x_n + \frac{h}{2}, y_n + \frac{k_2}{2}\right)$$

$$k_4 = hf(x_n + h, y_n + k_3)$$

# 4. Analysis of Experimental Data

## 4.1 *Probability Distributions for Discrete Random Variables*

**Notation:**

$P(r) = f(r) \Rightarrow$ the probability distribution of random variable $r$ is $f(r)$

$\mu$ = mean value of $r$ $= \sum_{i=1}^{n} r_i f(r_i)$

$\sigma^2$ = variance of $r$ $= \sum_{i=1}^{n} r_i^2 f(r_i) - \mu^2$

$\sigma$ = standard deviation of $r$

$\binom{n}{r}$ = binomial coefficient $= \dfrac{n!}{(n-r)!r!} = \binom{n}{n-r}$

Evaluate using Pascal's Triangle:

| | $r=0$ | 1 | 2 | 3 | 4 | 5 | 6 | 7 | 8 | 9 | 10 |
|---|---|---|---|---|---|---|---|---|---|---|---|
| n = 0 | 1 | | | | | | | | | | |
| 1 | 1 | 1 | | | | | | | | | |
| 2 | 1 | 2 | 1 | | | | | | | | |
| 3 | 1 | 3 | 3 | 1 | | | | | | | |
| 4 | 1 | 4 | 6 | 4 | 1 | | | | | | |
| 5 | 1 | 5 | 10 | 10 | 5 | 1 | | | | | |
| 6 | 1 | 6 | 15 | 20 | 15 | 6 | 1 | | | | |
| 7 | 1 | 7 | 21 | 35 | 35 | 21 | 7 | 1 | | | |
| 8 | 1 | 8 | 28 | 56 | 70 | 56 | 28 | 8 | 1 | | |
| 9 | 1 | 9 | 36 | 84 | 126 | 126 | 84 | 36 | 9 | 1 | |
| 10 | 1 | 10 | 45 | 120 | 210 | 252 | 210 | 120 | 45 | 10 | 1 |

**Binomial Distribution**

$n$ = number of trials with constant probability $p$ of success in each
$r$ = number of successes

$$P(r) = \binom{n}{r} p^r (1-p)^{n-r} \qquad r = 0, 1, 2, \ldots n$$

$$\mu = np \qquad \sigma^2 = np(1-p)$$

**Poisson Distribution**

$\mu$ = mean rate of occurrence of an event
$r$ = number of events actually occurring in unit time

$$P(r) = e^{-\mu} \frac{\mu^r}{r!} \qquad r = 0, 1, \ldots$$

$$\sigma^2 = \mu$$

## 4.2 *Probability Distributions for Continuous Random Variables*

**Exponential Distribution**

probability density function $f(x) = \lambda e^{-\lambda x}$ $x \geq 0, \lambda > 0$

$\mu = 1/\lambda$ $\sigma^2 = 1/\lambda^2$

**Normal Distribution**

The *standardised* normal distribution $N(0, 1)$ has $\mu = 0$ and $\sigma = 1$. Its probability density function is:

$$\phi(z) = \frac{1}{\sqrt{2\pi}} e^{-\frac{1}{2}z^2}$$

Its cumulative distribution function $\Phi(z) = \int_{-\infty}^{z} \frac{1}{\sqrt{2\pi}} e^{-\frac{1}{2}t^2} dt$

is the probability that the random variable is observed to have a value $\leq z$ (the shaded area shown in the graph of $N(0,1)$ on page 4-3).

**Percentage Points of the Normal Distribution $N(0, 1)$**

| $\Phi(z)$ | % (1 tail) | % (2 tails) | $z$ |
|---|---|---|---|
| 0.9500 | 5.0 | 10 | 1.6449 |
| 0.9750 | 2.5 | 5 | 1.9600 |
| 0.9900 | 1.0 | 2 | 2.3263 |
| 0.9950 | 0.5 | 1 | 2.5758 |

**Cumulative Distribution Function for $N(0, 1)$**

| $z$ | $\Phi(z)$ | $z$ | $\Phi(z)$ | $z$ | $\Phi(z)$ |
|---|---|---|---|---|---|
| 0.0 | 0.5000 | 1.0 | 0.8413 | 2.0 | 0.9772 |
| 0.1 | 0.5398 | 1.1 | 0.8643 | 2.1 | 0.9821 |
| 0.2 | 0.5793 | 1.2 | 0.8849 | 2.2 | 0.9861 |
| 0.3 | 0.6179 | 1.3 | 0.9032 | 2.3 | 0.9893 |
| 0.4 | 0.6554 | 1.4 | 0.9192 | 2.4 | 0.9918 |
| 0.5 | 0.6915 | 1.5 | 0.9332 | 2.5 | 0.9938 |
| 0.6 | 0.7257 | 1.6 | 0.9452 | 2.6 | 0.9953 |
| 0.7 | 0.7580 | 1.7 | 0.9554 | 2.7 | 0.9965 |
| 0.8 | 0.7881 | 1.8 | 0.9641 | 2.8 | 0.9974 |
| 0.9 | 0.8159 | 1.9 | 0.9713 | 2.9 | 0.9981 |
| (For negative values of $z$, use $\Phi(-z) = 1 - \Phi(z)$) | | | | 3.0 | 0.9987 |
| | | | | 4.0 | >0.9999 |

The *general* normal distribution $N(\mu, \sigma^2)$ has probability density function

$$f(x) = \frac{1}{\sigma\sqrt{2\pi}} e^{-\frac{(x-\mu)^2}{2\sigma^2}} = \phi\left(\frac{x-\mu}{\sigma}\right), \text{ where } \int_{-\infty}^{\infty} f(x) dx = 1$$

Its cumulative distribution $F(x) = \int_{-\infty}^{(x-\mu)/\sigma} \frac{1}{\sqrt{2\pi}} e^{-\frac{1}{2}t^2} dt = \Phi\left(\frac{x-\mu}{\sigma}\right)$.

Tables of $\phi(z)$ and $\Phi(z)$ may be used for $f(x)$ and $F(x)$ by taking $z = \frac{x-\mu}{\sigma}$.

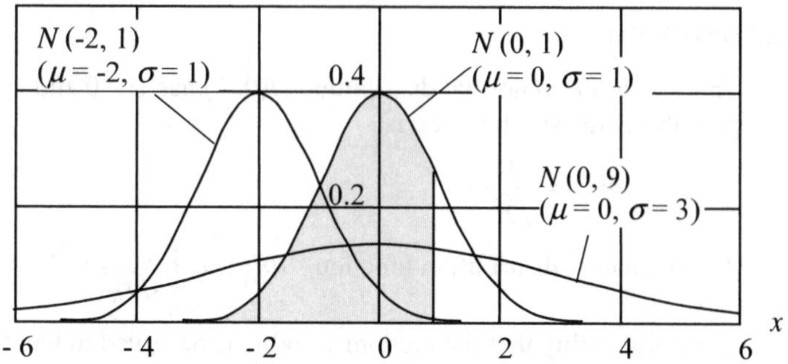


*Examples of normal distributions*

## 4.3 *Experimental Samples*

$x_1, x_2, \ldots x_n$ denote a set of $n$ observations of a random variable having a normal distribution whose mean $\mu$ is unknown.

| | |
|---|---|
| Range: | $x_{max} - x_{min}$ |
| Sample mean: | $m = \frac{1}{n}\sum x_i$ |
| Average deviation: | $\frac{1}{n}\sum \lvert x_i - m \rvert$ |
| Sample variance: | $s^2 = \frac{1}{n-1}\sum (x_i - m)^2$ |
| Sample Standard Deviation: | $s$ |
| Distribution of $x$: | $N(\mu, \sigma^2)$ |
| Distribution of $m$: | $N(\mu, \sigma^2/n)$ |
| Distribution of $\frac{m-\mu}{(\sigma/\sqrt{n})}$ | $N(0, 1)$ |
| Standard error of sample means: | $S = \frac{\sigma}{\sqrt{n}}$ |

If the population variance $\sigma^2$ is known,

95% confidence interval for $\mu$ *is* $m \pm 1.96\frac{\sigma}{\sqrt{n}}$

99% confidence interval for μ *is* $m \pm 2.58 \dfrac{\sigma}{\sqrt{n}}$

<u>If the population variance $\sigma^2$ is unknown,</u> $\dfrac{m-\mu}{s/\sqrt{n}}$ has the *t* distribution with $n-1$ degrees of freedom ($t_n$-1), and the 95% confidence interval for $\mu$ is obtained from $m \pm t_c \dfrac{s}{\sqrt{n}}$ and the table.

| 95% points of the *t*-distribution | | | | | |
|---|---|---|---|---|---|
| $n$ - 1 | $t_c$ | $n$ - 1 | $t_c$ | $n$ - 1 | $t_c$ |
| 1 | 12.7 | 6 | 2.45 | 12 | 2.18 |
| 2 | 4.30 | 7 | 2.36 | 15 | 2.13 |
| 3 | 3.18 | 8 | 2.31 | 20 | 2.09 |
| 4 | 2.78 | 9 | 2.26 | 30 | 2.04 |
| 5 | 2.57 | 10 | 2.23 | 60 | 2.00 |
| | | | | ∞ | 1.96 |

Thus, for $n > 20$, $m \pm 1.96\, s/\sqrt{n}$ is a good approximation to the population mean with a 95% confidence.

## 4.4 *Combination of Errors*

If results are normally distributed, the Most Probable Error $S_z$ in the calculated result $z = f(x, y, \text{etc})$ due to independent standard errors $S_x$, $S_y$, etc in $x$, $y$, etc is given by:

$$(S_z)^2 = \left(\frac{\partial z}{\partial x} S_x\right)^2 + \left(\frac{\partial z}{\partial y} S_y\right)^2 + \ldots \text{etc}$$

If the function *f*( ) consist of multiplied and divided terms *only* (i.e. no addition or subtraction):

$$\left(\frac{S_z}{z}\right)^2 = \left(n\frac{S_x}{x}\right)^2 + \left(m\frac{S_y}{y}\right)^2 + \ldots \text{etc}$$

where *n*, *m* etc are the powers of $x$, $y$ etc in *f*( ).

**Notes**:

(a) The maximum possible error ($S_z = \dfrac{\partial z}{\partial x} S_x + \dfrac{\partial z}{\partial y} S_y + \ldots \text{etc}$) is rarely of interest in engineering.

(b) Instrument 'rounding off' error $\pm\delta x$ may be treated as a normally distributed error by the equivalence $S_x \approx \frac{2}{3}\delta x$.

# 5. Mechanics

## 5.1 *Second Moment of Area*

**Definition:**

$$I_{XX} = \iint (y - y_1)^2 \, dx \, dy$$

The double integral is taken over the whole area of the shape. Similar expressions apply to axes in other directions.

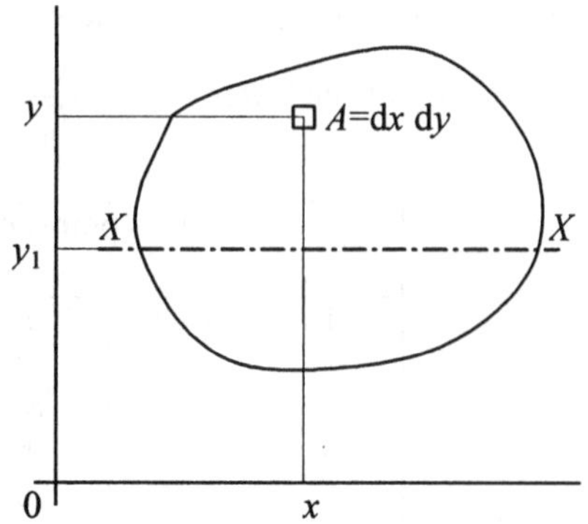


Radius of gyration $k$ is defined by $I = A k^2$ for a specified axis.

**Parallel axis theorem:**

$$I_{XX} = I_{GG} + A\,{y_0}^2$$

where $A$ is the area of the shape and $y_0$ is the perpendicular distance between an axis through the centroid, $GG$, and another parallel axis $XX$. $I_{GG}$ is the minimum second moment of area for all axes parallel to $GG$.

**Radius of Gyration of selected shapes**

| Shape | $k^2_{XX}$ | $k^2_{YY}$ | $A$ |
|---|---|---|---|
| Rectangle 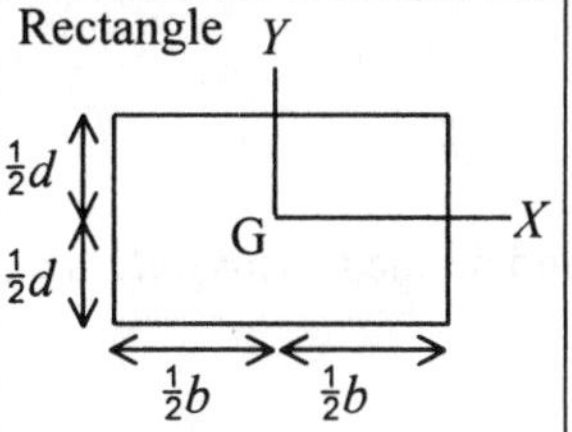  | $\frac{1}{12}d^2$ | $\frac{1}{12}b^2$ | $db$ |
| Circle 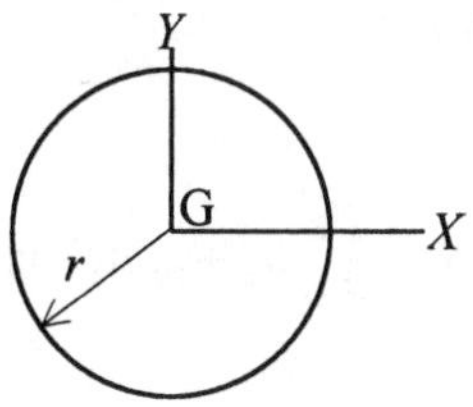  | $\frac{1}{4}r^2$ | $\frac{1}{4}r^2$ | $\pi r^2$ |

| Shape | $k^2_{XX}$ | $k^2_{YY}$ | $A$ |
|---|---|---|---|
| Ring 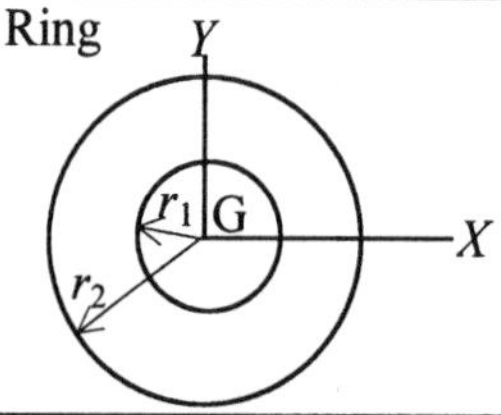  | $\frac{1}{4}(r_2^2 + r_1^2)$ | $\frac{1}{4}(r_2^2 + r_1^2)$ | $\pi(r_2^2 - r_1^2)$ |
| Semi-Circle 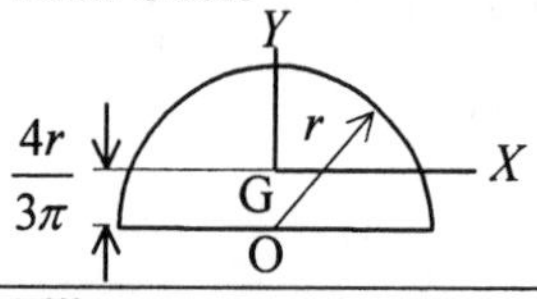  | $r^2\left\lvert\frac{1}{4} - \left(\frac{4}{3\pi}\right)^2\right\rvert$ <br> $k^2_{OO} = \frac{1}{4}r^2$ | $\frac{1}{4}r^2$ | $\frac{1}{2}\pi r^2$ |
| Ellipse 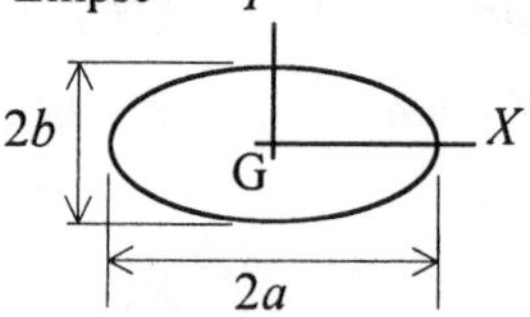  | $\frac{1}{4}b^2$ | $\frac{1}{4}a^2$ | $\pi ab$ |
| Parabola 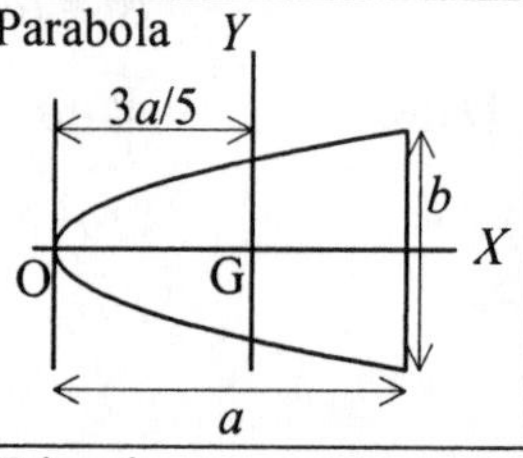  | $\frac{1}{20}b^2$ | $\frac{12}{175}a^2$ <br> $k^2_{OO} = \frac{3}{7}a^2$ | $\frac{2}{3}ab$ |
| Triangle 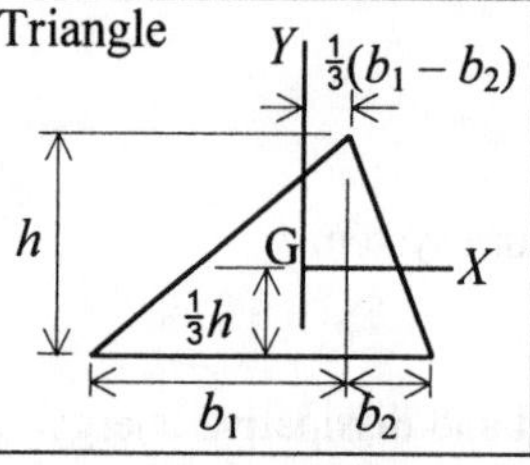  | $\frac{1}{18}h^2$ | $\frac{1}{18}(b_1^2 + b_1 b_2 + b_2^2)$ | $\frac{h}{2}(b_1 + b_2)$ |
| | $k^2_{XY} = \frac{1}{36}h(b_1 - b_2)$ | | |

**Mohr's Circle for Second Moment of Area**

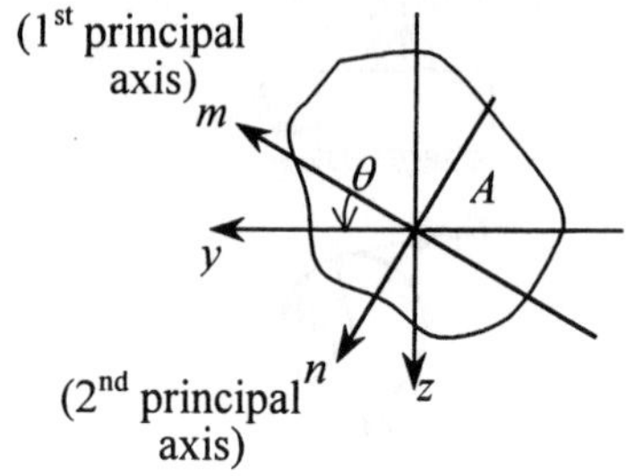


$$I_{yy} = \int_A z^2 \mathrm{d}A$$

$$I_{zz} = \int_A y^2 \mathrm{d}A$$

$$I_{yz} = I_{zy} = \int_A yz\,\mathrm{d}A$$

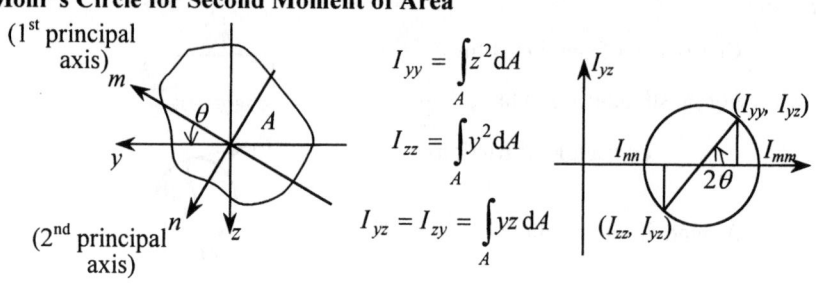

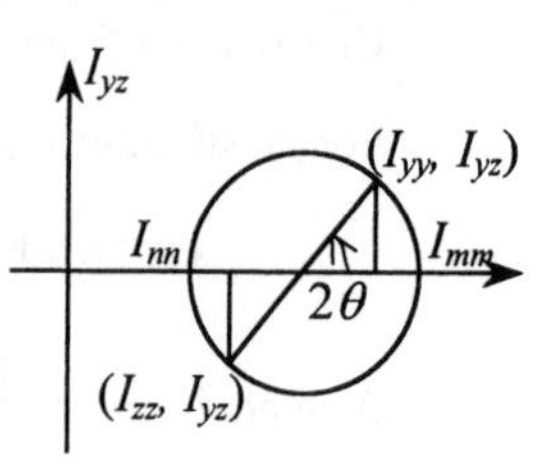

## 5.2 *Kinematics and Dynamics*

**Constant Acceleration Equations**

$$v = u + at \qquad v^2 = u^2 + 2ax$$

$$x = ut + \tfrac{1}{2}at^2$$

**Accelerations due to Rotation**

Central (centripetal) $= \underline{\omega} \times (\underline{\omega} \times \underline{r}) = r\omega^2 = \dfrac{v^2}{r}$

Coriolis $= 2\underline{\omega} \times \dfrac{d\underline{r}}{dt}$

**Force, Work and Energy**

| | *Linear* | *Rotational* |
|---|---|---|
| Newton's 2nd Law | $F = ma$ | $T = I\ddot{\theta} = I\dot{\omega}$ |
| Momentum, Impulse = $\int F dt$ | $mv$ | |
| Angular Momentum = $\int T dt$ | | $I\dot{\theta} = I\omega$ |
| Work, energy | $\int F dx$ | $\int T d\theta$ |
| Kinetic Energy | $\frac{1}{2} mv^2$ | $\frac{1}{2} I\dot{\theta}^2 = \frac{1}{2} I\omega^2$ |
| Power | $Fv$ | $T\dot{\theta}$ |

**Lagrange's Equation**

$$\frac{d}{dt}\left\{\frac{\partial \text{K.E.}}{\partial \dot{q}_i}\right\} - \frac{\partial \text{K.E.}}{\partial q_i} + \frac{\partial \text{P.E.}}{\partial q_i} + \frac{\partial \text{D.E.}}{\partial q_i} = Q_i$$

$i = 1,2,3$ for a 3 degree of freedom system.
$q_i$ is a generalised co-ordinate
$Q_i$ is a generalised force
K.E., P.E., D.E. are kinetic, potential and dissipative energy.

**Friction:**

Coefficient of static friction:

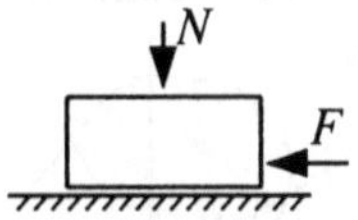


For no slipping: $\mu = \tan\phi \geq \dfrac{F}{N}$

($\phi$ = angle of friction)

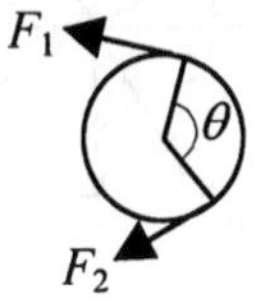


Around drum or pulley: $\dfrac{F_1}{F_2} = e^{\mu\theta}$

*Approximate Values of Friction Coefficients:*

| Material Combination | $\mu$ Static | | $\mu$ Dynamic | |
|---|---|---|---|---|
| | Dry | Lubricated | Dry | Lubricated |
| Steel/cast iron | 0.2 | 0.1 | 0.18 | 0.06 |
| Leather/metal | 0.6 | - | 0.48 | 0.15 |
| Brake or clutch lining/steel | - | - | 0.5 | - |
| Rubber/ashphalt | 0.8 | - | 0.5 | - |
| Filled PTFE/steel | - | - | 0.05-0.3 | - |

## 5.3 *Moments of Inertia*

**Definitions**:

Moment of Inertia: $I_{XX} = \int_M r^2 \mathrm{d}m$

$r$ is the perpendicular distance of the element of mass d$m$ from the axis $XX$; the integral is taken over the whole mass of the body.

Radius of gyration $k$ is defined by $I = Mk^2$, for a specified axis.

**Parallel axis theorem**:

$I_{XX} = I_{GG} + M y_0^2$

$M$ is the mass of the body, $y_0$ is the perpendicular distance between an axis $GG$ through the centre of mass and another axis $XX$ parallel to $GG$. $I_{GG}$ is the minimum moment of inertia for all axes in the given direction.

**Moment of inertia of a lamina**

For a lamina with uniform mass distribution over its area, moment of inertia about any axis in its plane is:

(mass per unit area)×(second moment of area about same axis).

***Perpendicular Axes Theorem***

$$J = I_{ZZ} = I_{XX} + I_{YY}$$

$I_{XX}$ and $I_{YY}$ are the moments of inertia of a lamina about two perpendicular axes in its plane, $J$ is the Polar Moment of Inertia, equal to the moment of inertia $I_{ZZ}$ about a third axis $ZZ$, perpendicular to and intersecting both $XX$ and $YY$.

**Moment of Inertia of a Uniform Density Body**

Axially-symmetric bodies may be treated as a succession of laminae.

| Shape | $k_{XX}^2$ | $k_{YY}^2$ | Volume |
|---|---|---|---|
| Rod 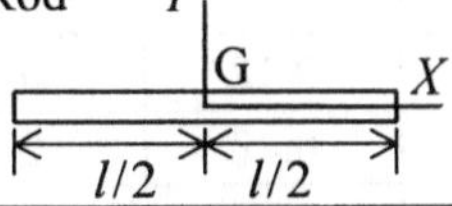  | - | $\frac{1}{12}l^2$ | - |
| Solid Cylinder 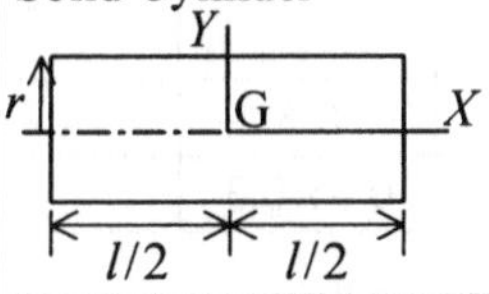  | $\frac{1}{2}r^2$ | $\frac{1}{4}r^2 + \frac{1}{12}l^2$ | $\pi r^2 l$ |
| Thin-walled Cylinder 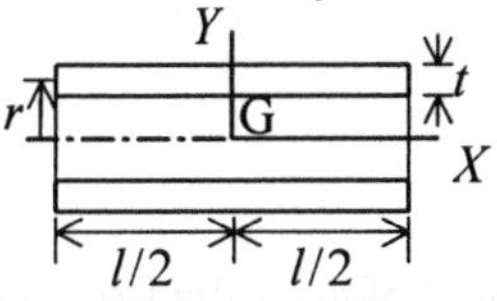  | $r^2 + \frac{1}{4}t^2$ | $\frac{1}{2}r^2 + \frac{1}{8}t^2 + \frac{1}{12}l^2$ | $2\pi rtl$ |
| Thick-walled Cylinder 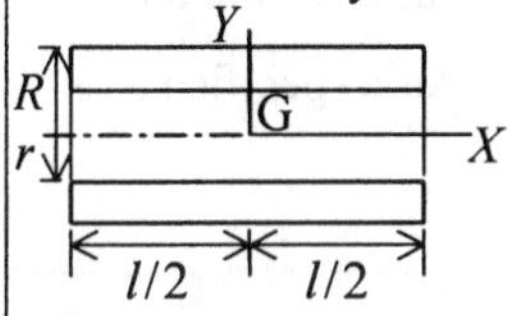  | $\frac{1}{2}(R^2 + r^2)$ | $\frac{1}{12}l^2 + \frac{1}{4}(R^2 + r^2)$ | $\pi(R^2 - r^2)l$ |
| Sphere 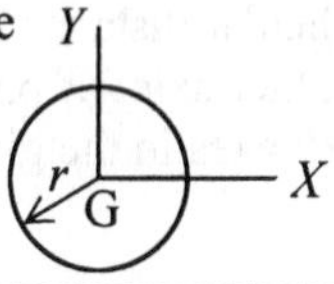  | $\frac{2}{5}r^2$ | $\frac{2}{5}r^2$ | $\frac{4}{3}\pi r^3$ |
| Cone 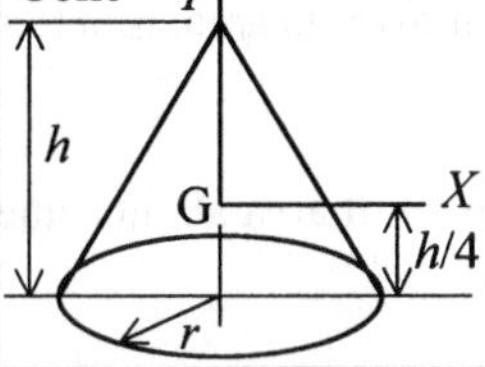  | $\frac{3}{80}(4r^2 + h^2)$ | $\frac{3}{10}r^2$ | $\frac{1}{3}\pi r^2 h$ |
| Rectangular Block 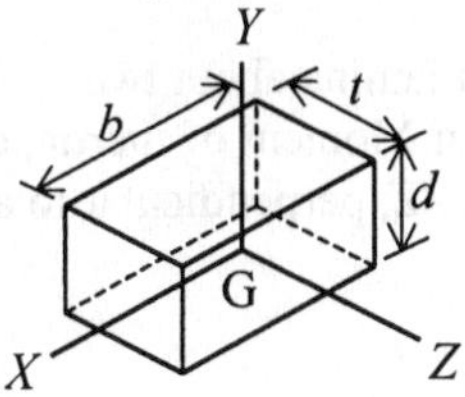  | $\frac{1}{12}(d^2 + t^2)$ | $\frac{1}{12}(b^2 + t^2)$ | $bdt$ |

# 6. Properties and Mechanics of Solids

## 6.1 *Bonding*

Condon-Morse Equation $V_{\text{total}} = -\frac{Ae^2}{r^n} + \frac{B}{r^m} + C$

Ionic Bond Equation $V_0 = -\frac{Z_1^* Z_2^* e^2}{4\pi d\varepsilon_0}(n - \frac{1}{n}) + \Delta E$

Theoretical Density $\rho = \frac{nA}{VN}$

## 6.2 *Atomic sizes in substitutional alloys*

| Element | Seitz radius $r_0$ (nm) at 20°C | Effective Valency in solution | Element | Seitz radius $r_0$ (nm) at 20°C | Effective Valency in solution |
|---|---|---|---|---|---|
| Al | 0.158 | 3 | P | 0.158 | 3 |
| Au | 0.159 | 1 | Pb | 0.195 | 4 |
| Cu | 0.141 | 1 | Si | 0.167 | 4 |
| Fe(α) | 0.141 | ? | Sn | 0.186 | 4 |
| Mg | 0.185 | 2 | Zn | 0.154 | 2 |
| Ni | 0.138 | 1 | | | |

## 6.3 *Phase Transformations*

Length and volume changes may be related by:

$$(1 + \Delta V/V) = (1 + \Delta L/L)^3$$

## 6.4 *Crystallography*

In the Miller index system:

| | |
|---|---|
| Specific plane | (h.k.l) |
| Family of planes | {h.k.l} |
| Specific direction | [h.k.l] |
| Family of directions | <h.k.l> |

Inter-planar spacing for cubics: $d_{(\text{h.k.l})} = \frac{a}{\sqrt{h^2 + k^2 + l^2}} = \frac{a}{\sqrt{N}}$

Quadratic form of Miller Indices for Cubic Structures

| | *N values* |
|---|---|
| Simple | 1,2,3,4,5,6,7,8,9,10,11,12,13,14,15,16,17,18,19,20,… |
| Face centred | 3,4,8,11,12,16,19,20,24,27,32,… |
| Body centred | 2,4,6,8,10,12,14,16,18,20,22,24,26,30,… |
| Diamond | 3,8,11,16,19,24,27,32,… |

## 6.5 *Defects and Diffusion Data*

Number of defects $n = \alpha N' e^{-\frac{Q}{kT}}$

Diffusivity $D = D_0 e^{-\frac{Q}{kT}}$

(Note: these equations may be expressed in terms of $R_0$ rather than $k$, but the value of $Q$ must be in the appropriate units.)

Macroscopic Diffusion:

(i) $D$ constant with composition, $\frac{dC}{dx}$ constant with time

(special case of (ii) below):

$$J = -D\frac{dC}{dx}$$

(ii) $D$ constant with composition, $\frac{dC}{dx}$ varies with time:

$$\frac{dC}{dt} = D\frac{d^2C}{dx^2}$$

Solution for case (ii) with constant surface concentration and impermeable sides:

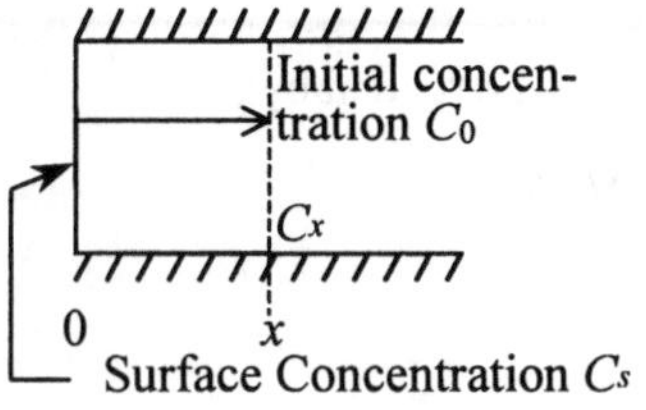


$$\frac{C_x - C_0}{C_s - C_0} = 1 - \text{erf}\left(\frac{x}{2\sqrt{Dt}}\right)$$

(Values of the error function erf( ) are tabulated on page 3.18.)

## 6.6 ***Fracture***

**Fracture Toughness:**

General Stress Intensity $K_1 = \sigma\sqrt{a}\,\gamma(\frac{a}{W})$

*Crack configurations with corresponding values of* $\gamma(\frac{a}{W})$

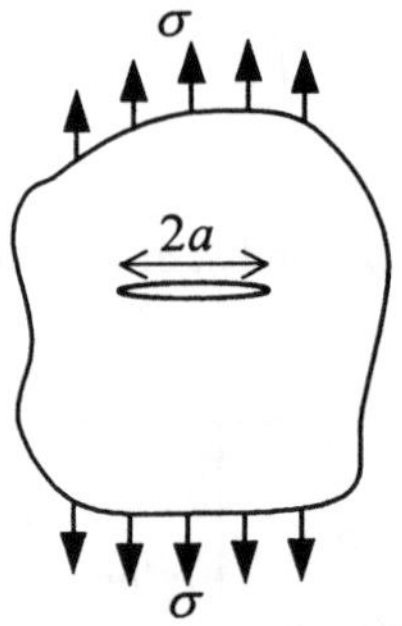


Isolated crack in an infinite plate:

$$\gamma(\frac{a}{W}) = \sqrt{\pi}$$

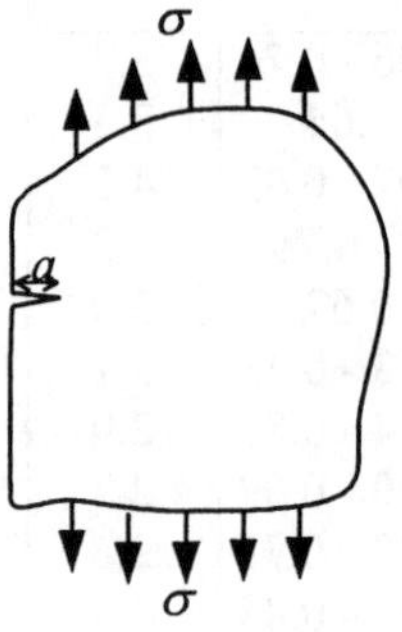


Surface crack in a semi-infinite plate:

$$\gamma(\frac{a}{W}) = 1.12\sqrt{\pi}$$

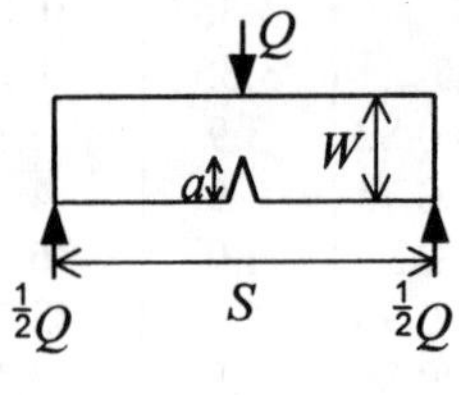


Surface crack in a three-point bend specimen:

$$\sigma = \frac{6M}{tW^2}, \quad M = \frac{QS}{4}$$

where $t$ is the plate thickness

$$\gamma(\frac{a}{W}) = 1.93 - 3.07\left(\frac{a}{W}\right) + 14.53\left(\frac{a}{W}\right)^2 - 25.11\left(\frac{a}{W}\right)^3 + 25.8\left(\frac{a}{W}\right)^4$$

**Fatigue:**

Manson-Coffin Law $\sqrt{N_p}\varepsilon_p = c = \text{constant}$

Miner's Rule $\sum\left(\frac{n_i}{N_i}\right) = 1$

Rayleigh Distribution $P(\sigma) = \sigma r^{-2} \exp\left\{-\frac{1}{2}(\frac{\sigma}{r})^2\right\}$

Fraction of peaks exceeding stress ($\sigma$) expressed in terms of $E$ $E(\sigma) = \exp\left\{-\frac{1}{2}(\frac{\sigma}{r})^2\right\}$

Paris Equation $\frac{\mathrm{d}a}{\mathrm{d}N} = A(\Delta K)^m = A_1 a^{m/2}$ ($A$, $A_1$ and $m$ are constants)

**Fatigue crack growth data for various materials**

| Material | $R$ $\sigma_{min}/\sigma_{max}$ | $m$ | $\Delta K$ for $\frac{\mathrm{d}a}{\mathrm{d}N} =$ $10^{-6}$ mm/c ($MN/m^{3/2}$) |
|---|---|---|---|
| Mild steel | 0.06 - 0.74 | 3.3 | 6.2 |
| in brine | 0.64 | 3.3 | 6.2 |
| cold rolled | 0.07 - 0.43 | 4.2 | 7.2 |
| Low alloy steel | 0 - 0.75 | 3.3 | 5.1 |
| Maraging steel | 0.67 | 3.0 | 3.5 |
| Austenitic steel 18/8 | 0.33 - 0.43 | 3.1 | 6.3 |
| Aluminium | 0.14 - 0.87 | 2.9 | 2.9 |
| Aluminium alloy 5083 | 0.20 - 0.69 | 2.7 | 1.6 |
| Aluminium alloy 6082 1% Mg, 1% Si, 0.7% Mn | 0.20 - 0.71 | 2.6 | 1.9 |
| Aluminium alloy L73 4.5% Cu | 0.50 - 0.88 | 4.4 | 2.1 |
| Aluminium alloy DTD 687A 5.5% Zn | 0.20 - 0.45 | 3.7 | 1.75 |
| Magnesium alloy ZW1 0.5% Zr | 0 | 3.35 | 0.94 |
| Magnesium alloy AM503 1.5% Mn | 0.5 | 3.35 | 0.69 |
| Copper | 0.07 - 0.82 | 3.9 | 4.3 |
| Phosphor bronze | 0.33 - 0.74 | 3.9 | 4.3 |
| Brass 60/40 | 0 - 0.33 | 4.0 | 6.3 |
| Titanium | 0.08 - 0.94 | 4.4 | 3.1 |
| Titanium alloy 5%Al | 0.17 - 0.86 | 3.8 | 3.4 |
| Titanium alloy 15% Mo | 0.28 - 0.71 | 3.5 | 3.0 |
| Nickel | 0 - 0.71 | 4.0 | 8.8 |
| Monel | 0 - 0.67 | 4.0 | 6.2 |
| Inconel | 0 - 0.71 | 4.0 | 8.2 |

## 6.7 *Corrosion Potentials*

**Standard oxidation/reduction (redox) potentials**

at 25°C, against normal hydrogen electrode

| | volts |
|---|---|
| $Au = Au^{3+} + 3e$ | +1.498 |
| $O_2 + 4H^{+} + 4e = 2H_2O$ | +1.229 |
| $Pt = Pt^{2+} + 2e$ | +1.2 |
| $Pd = Pd^{2+} + 2e$ | +0.987 |
| $Ag = Ag^{+} + e$ | +0.799 |
| $2Hg = Hg_2^{2+} + 2e$ | +0.788 |
| $Fe^{3+} + e = Fe^{2+}$ | +0.771 |
| $O_2 + 2H_2O + 4e = 4\,OH^{-}$ | +0.401 |
| $Cu = Cu^{2+} + 2e$ | +0.337 |
| $Sn^{4+} + 2e = Sn^{2+}$ | +0.15 |
| $2H^{+} + 2e = H_2$ | 0.000 |
| $Pb = Pb^{2+} + 2e$ | -0.126 |
| $Sn = Sn^{2+} + 2e$ | -0.136 |
| $Ni = Ni^{2+} + 2e$ | -0.250 |
| $Co = Co^{2+} + 2e$ | -0.277 |
| $Cd = Cd^{2+} + 2e$ | -0.403 |
| $Fe = Fe^{2+} + 2e$ | -0.440 |
| $Cr = Cr^{3+} + 3e$ | -0.744 |
| $Zn = Zn^{2+} + 2e$ | -0.763 |
| $Al = Al^{3+} + 3e$ | -1.662 |
| $Mg = Mg^{2+} + 2e$ | -2.363 |
| $Na = Na^{+} + e$ | -2.714 |
| $K = K^{+} + e$ | -2.925 |

Passive (Cathodic)

Active (Anodic)

| **Galvanic Series in Sea Water** |
|---|
| Titanium |
| Monel |
| Passive 18/8 |
| Silver |
| Nickel |
| Cupro-nickel |
| Aluminium bronze |
| Copper |
| Brass |
| Active 18/8 |
| Cast iron |
| Steel |
| Aluminium |
| Zinc |
| Magnesium |

## 6.8 *Structure and Properties of some Elements*

| | Metals | | | Semiconductors | |
|---|---|---|---|---|---|
| | Copper Cu | Iron Fe | Aluminium Al | Germanium Ge | Silicon Si |
| Crystal Structure | f.c.c. | b.c.c. | f.c.c. | diamond | diamond |
| Bonding | metallic | metallic | metallic | covalent | covalent |
| Lattice constant (nm) | 0.3615 | 0.2866 | 0.4049 | 0.5658 | 0.5431 |
| Atomic volume ($nm^3$) | $11.81\times10^{-3}$ | $11.77\times10^{-3}$ | $16.6\times10^{-3}$ | $22.64\times10^{-3}$ | $20.02\times10^{-3}$ |
| Density ($kg/m^3$) | 8960 | 7870 | 2690 | 5320 | 2330 |
| Melting point (°C) | 1083 | 1530 | 660 | 958.5 | 1412 |
| $\alpha$ (μm/(mK)) | 16.7 | 12.1 | 23.5 | 5.75 | 7.6 |
| Thermal conductivity at 20°C (W/mK) | 410 | 81 | 238 | 63 | 156 |
| Cohesive Energy (J/kg mol) | $3.38\times10^8$ | $4.05\times10^8$ | - | $3.72\times10^8$ | $4.39\times10^8$ |
| Resisivity at 20°C (Ωm) | $1.72\times10^{-8}$ | $10\times10^{-8}$ | $2.69\times10^{-8}$ | - | - |
| Temperature coefficient of resistance ($K^{-1}$) | +0.0043 | +0.0065 | +0.0042 | - | - |
| Ionisation Potential (eV) | 7.72 | 7.87 | 5.99 | - | - |
| Work function (eV) | 4.51 | 4.60 | 4.19 | - | - |
| Mobility ($m^2/(Vs)$)<br>electrons<br>holes | - | - | - | 0.38<br>0.18 | 0.19<br>0.05 |
| Energy gap (room temperature) | - | - | - | 0.67 | 1.107 |
| Density of states effective mass<br>electrons<br>holes | - | - | - | $0.35\ m_e$<br>$0.56\ m_e$ | $0.58\ m_e$<br>$1.06\ m_e$ |

| 1s | | |
|---|---|---|
| | 1 H | 2 He |
| | s<br>1 | s<br>2 |

| | IA | IIA | | IIIA | IVA | VA | VIA | VIIA | ← VIII | VIII | VIII → | IB | IIB | | IIIB | IVB | VB | VIB | VIIB | |
|---|---|---|---|---|---|---|---|---|---|---|---|---|---|---|---|---|---|---|---|---|
| 2s | 3 Li | 4 Be | | | | | | | | | | | | 2p | 5 B | 6 C | 7 *N* | 8 *O* | 9 *F* | 10 *Ne* |
| 3s | 11 Na | 12 Mg | | | | | | | | | | | | 3p | 13 Al | 14 Si | 15 P | 16 S | 17 Cl | 18 Ar |
| 4s | 19 K | 20 Ca | 3d | 21 Sc | 22 Ti | 23 V | 24 Cr | 25 Mn | 26 Fe | 27 Co | 28 Ni | 29 Cu | 30 Zn | 4p | 31 Ga | 32 Ge | 33 As | 34 Se | 35 Br | 36 Kr |
| 5s | 37 Rb | 38 Sr | 4d | 39 Y | 40 Zr | 41 Nb | 42 Mo | 43 Tc | 44 Ru | 45 Rh | 46 Pd | 47 Ag | 48 Cd | 5p | 49 *In* | 50 Sn | 51 Sb | 52 Te | 53 I | 54 Xe |
| 6s | 55 Cs | 56 Ba | 5d | 57 La | 72 Hf | 73 Ta | 74 W | 75 Re | 76 Os | 77 Ir | 78 Pt | 79 Au | 80 Hg | 6p | 81 Tl | 82 Pb | 83 Bi | 84 Po | 85 At | 86 Rn |
| | | | | | d s<br>2 2 | d s<br>3 2<br>4 1<br>3 2 | d s<br>5 1<br>5 1<br>4 2 | d s<br>5 2<br>6 1<br>5 2 | d s<br>6 2<br>7 1<br>6 2 | d s<br>7 2<br>8 1<br>7 2 | d s<br>8 2<br>10 0<br>9 1 | d s<br>10 1 | d s<br>10 2 | | s p<br>2 1 | s p<br>2 2 | s p<br>2 3 | s p<br>2 4 | s p<br>2 5 | s p<br>2 6 |
| 7s | 87 Fr | 88 Ra | 6d | 89 Ac | | | | | | | | | | | | | | | | |
| | s<br>1 | s<br>2 | | d s<br>1 2 | | | | | | | | | | | | | | | | |

| 4f | 58 Ce | 59 Pr | 60 Nd | 61 Pm | 62 Sm | 63 Eu | 64 Gd | 65 Tb | 66 Dy | 67 Ho | 68 Er | 69 Tm | 70 Yb | 71 Lu |
|---|---|---|---|---|---|---|---|---|---|---|---|---|---|---|
| 4 f<br>5 s<br>5 p<br>5 d<br>6 s | 2<br>2<br>6<br>0<br>2 | 3 | 4 | 5 | 6 | 7 | f 7<br>s 2<br>p 6<br>d 1<br>s 2 | 9<br>2<br>6<br>0<br>2 | 10 | 11 | 12 | 13 | 14 | f 14<br>s 2<br>p 6<br>d 1<br>s 2 |
| 5f | 90 Th | 91 Pa | 92 U | 93 Np | 94 Pu | 95 Am | 96 Cm | 97 Bk | 98 Cf | 99 Es | 100 Fm | 101 Md | 102 No | 103 Lw |
| 5 f<br>6 s<br>6 p<br>6 d<br>7 s | 0<br>2<br>6<br>2<br>2 | 2<br>2<br>6<br>1<br>2 | 3 | 4 | 5 | 6 | 7 | 8 | 9 | 10 | 11 | 12 | 13 | 14 |

←Outer sub-shells as for Ce→ (Pr to Eu)

Outer sub-shells as for Ce & Tb (Dy to Yb)

← Outer sub-shells as for Pa → (U to Lw)

} IIIA (La → 4f, Ac → 5f)

| ⊕ | 1 | 2 | 3 | 4 | 5 | 6 | 7 | ← Principal Shell |
|---|---|---|---|---|---|---|---|---|
| | s | s p | s p d | s p d f | s p d f | s p d | s | ← Sub-shell |
| | 2 | 2 6 | 2 6 10 | 2 6 10 14 | 2 6 10 14 | 2 6 10 | 2 | ← Number of electrons required to fill sub-shell |

## 6.10 ***The Elements in Alphabetical Order of Symbol***

| *Symbol* | *Name* | *Atomic Number* | *Atomic Weight* | *Symbol* | *Name* | *Atomic Number* | *Atomic Weight* |
|---|---|---|---|---|---|---|---|
| A or Ar | Argon | 18 | 39.948 | Mn | Manganese | 25 | 54.9380 |
| Ac | Actinium | 89 | - | Mo | Molybdenum | 42 | 95.94 |
| Ag | Silver | 47 | 107.870 | N | Nitrogen | 7 | 14.0067 |
| Al | Aluminium | 13 | 26.9815 | Na | Sodium | 11 | 22.9898 |
| Am | Americium | 95 | - | Nb | Niobium | 41 | 92.906 |
| As | Arsenic | 33 | 74.9216 | Nd | Neodymium | 60 | 144.24 |
| At | Astatine | 85 | - | Ne | Neon | 10 | 20.183 |
| Au | Gold | 79 | 196.967 | Ni | Nickel | 28 | 58.71 |
| B | Boron | 5 | 10.811 | No | Nobelium | 102 | - |
| | | | ±0.003 | Np | Neptunium | 93 | - |
| Ba | Barium | 56 | 137.34 | O | Oxygen | 8 | 15.9994 |
| Be | Beryllium | 4 | 9.0122 | | | | ±0.0001 |
| Bi | Bismuth | 83 | 208.980 | Os | Osmium | 76 | 190.2 |
| Bk | Berkelium | 97 | - | P | Phosphorus | 15 | 30.9738 |
| Br | Bromine | 35 | 79.909 | Pa | Protoactinium | 91 | - |
| C | Carbon | 6 | 12.01115 | Pb | Lead | 82 | 207.19 |
| | | | ±0.00005 | Pd | Palladium | 46 | 106.4 |
| Ca | Calcium | 20 | 40.08 | Pm | Promethium | 61 | - |
| Cd | Cadmium | 48 | 112.40 | Po | Polonium | 84 | - |
| Ce | Cerium | 58 | 140.12 | Pr | Praseodymium | 59 | 140.907 |
| Cf | Californium | 98 | - | Pt | Platinum | 78 | 195.09 |
| Cl | Chlorine | 17 | 35.453 | Pu | Plutonium | 94 | - |
| Cm | Curium | 96 | - | Ra | Radium | 88 | - |
| Co | Cobalt | 27 | 58.9332 | Rb | Rubidium | 37 | 85.47 |
| Cr | Chromium | 24 | 51.996 | Re | Rhenium | 75 | 186.2 |
| Cs | Caesium | 55 | 132.905 | Rh | Rhodium | 45 | 102.905 |
| Cu | Copper | 29 | 63.54 | Rn | Radon | 86 | - |
| Dy | Dysprosium | 66 | 162.50 | Ru | Ruthenium | 44 | 101.07 |
| Er | Erbium | 68 | 167.26 | S | Sulphur | 16 | 32.064 |
| Es | Einsteinium | 99 | - | | | | ±0.003 |
| Eu | Europium | 63 | 151.96 | Sb | Antimony | 51 | 121.75 |
| F | Fluorine | 9 | 18.9984 | Sc | Scandium | 21 | 44.956 |
| Fe | Iron | 26 | 55.847 | Se | Selenium | 34 | 78.96 |
| Fm | Fermium | 100 | - | Si | Silicon | 14 | 28.086 |
| Fr | Francium | 87 | - | | | | ±0.001 |
| Ga | Gallium | 31 | 69.72 | Sm | Samarium | 62 | 150.35 |
| Gd | Gadolinium | 64 | 157.25 | Sn | Tin | 50 | 118.69 |
| Ge | Germanium | 32 | 72.59 | Sr | Strontium | 38 | 87.62 |
| H | Hydrogen | 1 | 1.00797 | Ta | Tantalum | 73 | 180.948 |
| | | | ±0.00001 | Tb | Terbium | 65 | 158.924 |
| He | Helium | 2 | 4.0026 | Tc | Technetium | 43 | - |
| Hf | Hafnium | 72 | 178.49 | Te | Tellurium | 52 | 127.60 |
| Hg | Mercury | 80 | 200.59 | Th | Thorium | 90 | 232.038 |
| Ho | Holmium | 67 | 164.930 | Ti | Titanium | 22 | 47.90 |
| I | Iodine | 53 | 126.9044 | Tl | Thallium | 81 | 204.37 |
| In | Indium | 49 | 114.82 | Tm | Thulium | 69 | 168.934 |
| Ir | Iridium | 77 | 192.2 | U | Uranium | 92 | 238.03 |
| K | Potassium | 19 | 39.102 | V | Vanadium | 23 | 50.942 |
| Kr | Krypton | 36 | 83.80 | W | Tungsten | 74 | 183.85 |
| La | Lanthanum | 57 | 138.91 | Xe | Xenon | 54 | 131.30 |
| Li | Lithium | 3 | 6.939 | Y | Yttrium | 39 | 88.905 |
| Lu | Lutetium | 71 | 174.97 | Yb | Ytterbium | 70 | 173.04 |
| Md | Mendeleevium | 101 | - | Zn | Zinc | 30 | 65.37 |
| Mg | Magnesium | 12 | 24.312 | Zr | Zirconium | 40 | 91.22 |

Note: the values given normally indicate the mean atomic weight of the mixture of isotopes found in nature. Particular attention is drawn to the values for hydrogen, boron, carbon, oxygen, silicon and sulphur, where the deviation shown is due to variation in relative concentration of isotopes.

## 6.11 *Polymer Names and Structures*

| Common Name | Chemical Name | Abbreviation | Structure | State |
|---|---|---|---|---|
| Polythene | Polyethylene HD | HDPE | $\left[\!\!-CH_2\!-\!CH_2\!-\!\right]_n$ | Crystalline |
| PVC | Polyvinyl chloride | PVC | $\left[\!\!-CH_2\!-\!\underset{\underset{Cl}{\mid}}{CH}\!-\!\right]_n$ | Amorphous/ slightly crystalline |
| Polystyrene | Polystyrene | PS | $\left[\!\!-CH_2\!-\!\underset{C_6H_5}{CH}\!-\!\right]_n$ | Amorphous/ crystalline |
| Perspex | Polymethyl methacrylate | PMMA | $\left[\!\!-CH_2\!-\!\underset{\underset{O=C-O-CH_3}{\mid}}{\overset{\overset{CH_3}{\mid}}{C}}\!-\!\right]_n$ | Amorphous |
| PTFE | Polytetra-fluorethylene | PTFE | $\left[\!\!-CF_2\!-\!CF_2\!-\!\right]_n$ | Crystalline |
| Polypropylene | Polypropylene | PP | $\left[\!\!-CH_2\!-\!\underset{\underset{CH_3}{\mid}}{CH}\!-\!\right]_n$ | Amorphous/ crystalline |
| Nylon | Polyamide 6:6 | PA | $\left[\!\!-NH\text{-}(CH_2)_6\text{-}NH\text{-}\underset{\underset{O}{\parallel}}{C}\text{-}(CH_2)_4\text{-}\underset{\underset{O}{\parallel}}{C}-\right]_n$ | Crystalline |
| Rubber | Polyisoprene | PI | $\left[\!\!-CH_2\!-\!\underset{}{\overset{\overset{CH_3}{\mid}}{C}}\!=\!CH\!-\!CH_2\!-\!\right]_n$ | Elastomer |
| Polycarbonate | Polycarbonate | PC | $\left[\!\!-O\text{-}\underset{\underset{O}{\parallel}}{C}\text{-}O\text{-}\bigcirc\text{-}\underset{\underset{CH_3}{\mid}}{\overset{\overset{CH_3}{\mid}}{C}}\text{-}\bigcirc-\right]_n$ | Amorphous/ crystalline |
| Bakelite | Phenol formal-dehyde resin | PF | $\left[\!\!-\underset{\underset{CH_2}{\mid}}{\overset{\overset{OH}{\mid}}{\bigcirc}}\text{-}CH_2\text{-}\underset{\underset{CH_2}{\mid}}{\overset{\overset{OH}{\mid}}{\bigcirc}}\text{-}CH_2-\right]_n$ | Amorphous |

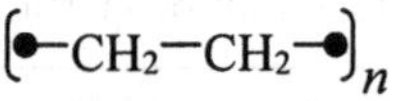

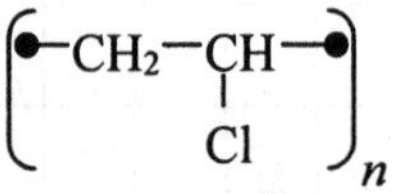

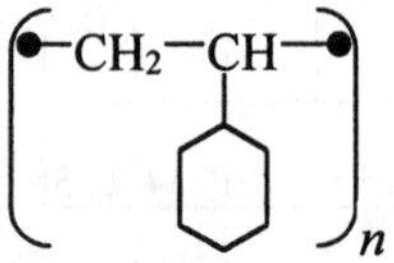

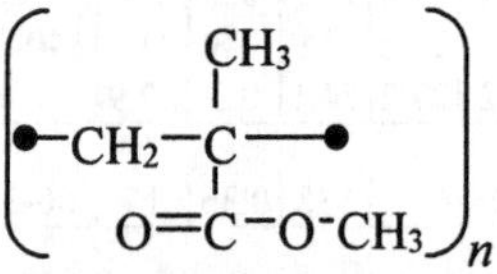

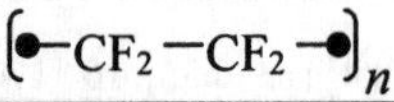

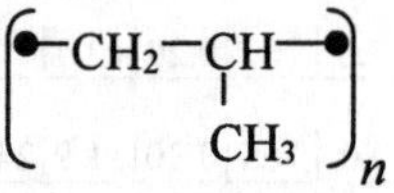

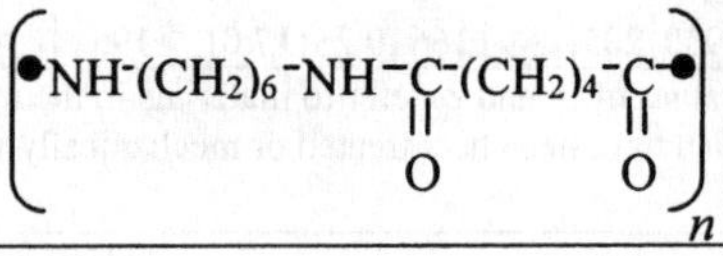

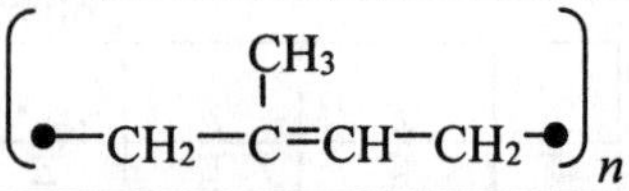

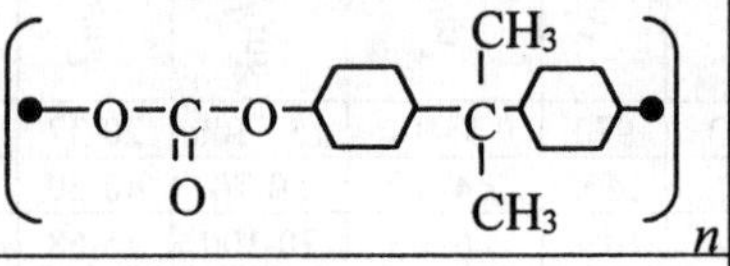

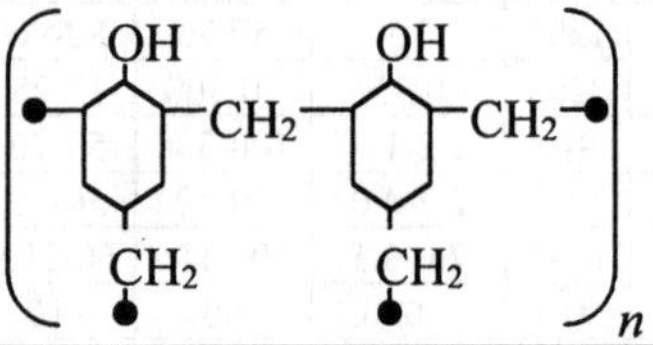

# 7. Properties of Materials

## 7.1 *Solids*

Unless otherwise stated, all properties are given for a temperature of 20°C.

**Metals**

| | $\rho$ | $E$ | $G$ | $K$ | $\nu$ | $\alpha$ | $\sigma_{y(0.2)}$ * | $\sigma_f$ * | $K_{IC}$ | Melting Range | $c_p$ | $k$ | Resistivity |
|---|---|---|---|---|---|---|---|---|---|---|---|---|---|
| | kg/m$^3$ | GN/m$^2$ | GN/m$^2$ | GN/m$^2$ | | µm/(mK) | MN/m$^2$ | MN/m$^2$ | MN/m$^{3/2}$ | °C | kJ/(kgK) | W/(mK) | Ωm |
| **Aluminium** | | | | | | | | | | | | | |
| 99.99% | 2700 | 69 | 25.5 | 75.5 | 0.34 | 23.5 | <25 | <58 | 45 | >660 | 0.9 | 244 | $2.7\times10^{-8}$ |
| **Aluminium Alloy** | | | | | | | | | | | | | |
| 2024 | 2770 | 74 | 27.6 | 77.1 | 0.3 | 23 | 325-345 | 470-480 | 10-50 | 500-640 | 0.88 | 151 | $5.7\times10^{-8}$ |
| 5083 | 2670 | 71 | 26.8 | 73.9 | 0.3 | 24.5 | 140-285 | 310-380 | 30-50 | 580-645 | 0.89 | 109 | $6.1\times10^{-8}$ |
| 7075TF | 2800 | 72 | 27.2 | 74.1 | 0.3 | 32.9 | <505 | <570 | 20-70 | 475-630 | 0.88 | 130 | $5.2\times10^{-8}$ |
| **Copper** | | | | | | | | | | | | | |
| 99.9% | 8900 | 125 | 48.3 | 138 | 0.35 | 17 | 60-325 | 220-385 | >100 | 1083 | 0.39 | 393 | $1.7\times10^{-8}$ |
| **Brass** | | | | | | | | | | | | | |
| 65/35 | 8450 | 102 | 37.3 | 115 | 0.35 | 21 | 290-300 | 460-480 | 30-100 | 900-920 | 0.38 | 125 | $6.2\times10^{-8}$ |
| **Cupro-Nickel** | | | | | | | | | | | | | |
| 70/30 | 8950 | 147 | 54.4 | 163 | 0.35 | 16 | 170-200 | 400-430 | >100 | 1170-1240 | 0.34 | 29 | $\sim20\times10^{-8}$ |
| **Mild Steel** | | | | | | | | | | | | | |
| 055M15 | 7860 | 207 | 82.2 | 169 | 0.29 | 12.4 | >150 | >310 | 140 | >1550 | 0.46 | 50 | $16.9\times10^{-8}$ |
| **0.4%C Steel** | | | | | | | | | | | | | |
| 080M40 | 7860 | 210 | 80 | 165 | 0.30 | 11.9 | 245-280 | 510-550 | 20-50 | >1550 | 0.46 | 50 | $17.1\times10^{-8}$ |
| **Stainless Steel** | | | | | | | | | | | | | |
| 304S15 | 7930 | 205 | 84 | 166 | 0.28 | 17.0 | >195 | >480 | >130 | >1550 | 0.46 | 15 | $70\times10^{-8}$ |

* the lower values of $\sigma_y$ and $\sigma_f$ refer to materials in the as-manufactured state, the higher values refer to materials which have been heat treated or mechanically worked.

**Polymers**

| | $\rho$ | $E$ | $\alpha$ | $\sigma_f$ * | $K_{IC}$ | Glass Tr. Temp. $T_g$ | $c_p$ | $k$ | Resis-tivity |
|---|---|---|---|---|---|---|---|---|---|
| | kg/m$^3$ | GN/m$^2$ | µm/(mK) | MN/m$^2$ | MN/m$^{3/2}$ | °C | kJ/(kgK) | W/(mK) | Ωm |
| Polythene HD | 970 | 0.55-1.0 | 150-300 | 20-37 | 2-5 | 300 | 2.1 | 0.52 | $>10^{14}$ |
| PVC | 1400 | 2.4-3.0 | 50-70 | 40-60 | 2 | 350 | | 0.15 | $>10^{12}$ |
| Polystyrene | 1050 | 3.0-3.3 | 70-100 | 35-68 | 2 | 370 | 1.3-1.5 | 0.1-0.15 | $>10^{15}$ |
| Perspex | 1200 | 3.3 | 50-70 | 80-90 | 1.5 | 378 | 1.5 | 0.2 | $>10^{14}$ |
| PTFE | 2200 | 0.35 | 70-100 | 17-28 | | 399 | 1.05 | 0.25 | $>10^{15}$ |
| Polypropylene | 910 | 1.2-1.7 | 100-300 | 50-70 | 3.5 | 253 | 1.9 | 0.2 | $>10^{13}$ |
| Polycarbonate | 1200 | 2.2-4.0 | 60-70 | 50-60 | 5-8 | 360 | 1.2-1.3 | 0.19 | $>10^{16}$ |
| Nylon | 1150 | 2.0-3.5 | 100-150 | 60-110 | 3-5 | 323 | 1.9 | 0.2-0.25 | $>10^{14}$ |
| Rubber | 910 | 0.002-0.1 | ~600 | ~10 | | 203 | ~2.5 | ~0.15 | $>10^{10}$ |
| Bakelite | 1300 | 8 | 20-60 | 35-55 | | - | 1.5-1.7 | 0.12-0.24 | $>10^{10}$ |

* $\sigma_f$ range is related to the exact composition of the polymer and the degree of drawing under stress.

| Other Materials | | | | | | | | | | | | |
|---|---|---|---|---|---|---|---|---|---|---|---|---|
| | $\rho$ | $E$ | $G$ | $K$ | $\nu$ | $\alpha$ | $\sigma_f$ | $K_{IC}$ | Melting | $c_p$ | $k$ | Resistivity |
| | kg/m$^3$ | GN/m$^2$ | GN/m$^2$ | GN/m$^2$ | | µm/(mK) | MN/m$^2$ | MN/m$^{3/2}$ | °C | kJ/(kgK) | W/(mK) | Ωm |
| Alumina | 3800 | 350 | | | 0.24 | 9 | 140-200 | 3-5 | >2000 | 1.05 | 29 | $>10^{12}$ |
| Quartz | 2650 | 73 | 31 | 37 | 0.17 | 0.5 | | ~0.5 | >1600 | 0.73 | 12 | $>10^{14}$ |
| Glass (soda) | 2480 | 74 | 29 | 41 | 0.22 | 8.5 | 30-90 | 0.7 | 1000 | 0.99 | 1 | $>10^{14}$ |
| Concrete | 2400 | 14 | | | 0.1-0.2 | 13 | ~20* | 0.2 | | 1.1 | 1.1 | |
| Firebrick | 2100 | | | | | 3-10 | | ~1.0 | | 0.81 | 0.4 | |
| Wood (along grain) | 400-800 | 8-13 | | | | 3-5 | 10-100 | 0.05-1.0 | | ~2.0 | ~0.15 | |

* failure stress in compression

## 7.2 ***Liquids***

| | $\rho$ | $c_p$ | $\mu$ | $k$ | $\sigma$ | $\beta$ | $K$ |
|---|---|---|---|---|---|---|---|
| | kg/m$^3$ | kJ/(kgK) | Ns/m$^2$ | W/(mK) | N/m | K$^{-1}$ | GN/m$^2$ |
| Water | 1000 | 4.19 | $1.002\times10^{-3}$ | 0.6 | 0.073 | $0.21\times10^{-3}$ | 2.3 |
| Mercury | 13 600 | 0.14 | $1.55\times10^{-3}$ | 8.7 | 0.51 | $0.18\times10^{-3}$ | |
| Castor oil | 960 | 2.20 | $3.14\times10^{-3}$ | 0.18 | 0.039 | | |
| Benzene | 880 | 1.80 | $0.656\times10^{-3}$ | 0.16 | 0.029 | $1.34\times10^{-3}$ | |
| Ethyl alcohol | 790 | 2.86 | $1.20\times10^{-3}$ | 0.19 | 0.022 | $1.08\times10^{-3}$ | |
| Engine oil (typical) | 890 | 1.9 | $80\times10^{-3}$ | 0.15 | | $0.8\times10^{-3}$ | |
| Freon 12 | 1350 | 0.96 | $0.273\times10^{-3}$ | 0.073 | | | |

## 7.3 ***Gases***

Properties at 20°C, 1 bar

| | $\rho$ | $c_p$ | $\mu$ | $k$ | $R$ | $\gamma = c_p/c_v$ |
|---|---|---|---|---|---|---|
| | kg/m$^3$ | kJ/(kgK) | Ns/m$^2$ | W/(mK) | kJ/(kgK) | |
| hydrogen | 0.082 | 14.3 | $0.88\times10^{-5}$ | 0.18 | 4.16 | 1.40 |
| helium | 0.164 | 5.23 | $1.96\times10^{-5}$ | 0.14 | 2.08 | 1.66 |
| nitrogen | 1.16 | 1.04 | $1.76\times10^{-5}$ | 0.026 | 0.294 | 1.40 |
| oxygen | 1.31 | 0.91 | $2.03\times10^{-5}$ | 0.026 | 0.260 | 1.40 |
| carbon dioxide | 1.8 | 0.84 | $1.47\times10^{-5}$ | 0.017 | 0.190 | 1.28 |
| air | 1.19 | 1.005 | $1.82\times10^{-5}$ | 0.026 | 0.287 | 1.40 |

### Composition of air

| Air (28.96)* | Nitrogen $N_2$ (28.013) | Oxygen $O_2$ (31.999) | Argon Ar (39.948) | Carbon dioxide $CO_2$ (44.010) |
|---|---|---|---|---|
| By volume | 0.7809 | 0.2095 | 0.0093 | 0.0003 |
| By mass | 0.7553 | 0.2314 | 0.0128 | 0.0005 |

* the numbers in brackets are the molecular weights of the gases.

### Critical Constants

| | Molecular weight | $T_c$ (K) | $P_c$ (bar) | $\rho_c$ (kg/m$^3$) |
|---|---|---|---|---|
| hydrogen | 2.02 | 33.3 | 13.0 | 31 |
| helium (4) | 4.00 | 5.3 | 2.29 | 69.3 |
| water vapour | 18.02 | 647.30 | 221.2 | 318.3 |
| nitrogen | 28.01 | 126.1 | 33.9 | 311 |
| oxygen | 32.00 | 154.4 | 50.4 | 430 |
| carbon dioxide | 44.01 | 304.15 | 73.8 | 468 |

## 7.4 *Fuels*

### Gases

| | Composition by volume % | | | | | | | Relative density (air = 1) | Calorific value at 1.01325 bar, 15°C MJ/m$^3$ | | Theoretical air |
|---|---|---|---|---|---|---|---|---|---|---|---|
| | $N_2$ | $H_2$ | $CH_4$ | $C_2H_6$ | $C_3H_8$ | $C_4H_{10}$ | $C_3H_6$ | | Gross | Net | vol/-vol |
| hydrogen | | 100 | | | | | | 0.0696 | 12.10 | 10.22 | 2.38 |
| methane | | | 100 | | | | | 0.5537 | 37.71 | 33.95 | 9.52 |
| North Sea gas | 1.5 | | 94.4 | 3.0 | 0.5 | 0.2 | | 0.589 | 38.62 | 34.82 | 9.75 |
| propane* | | | | 1.5 | 91.0 | 2.5 | 5.0 | 1.523 | 98.87 | 86.43 | 23.76 |
| butane* | | | 0.1 | 0.5 | 7.2 | 87.0 | 4.2 | 1.941 | 117.75 | 108.69 | 29.92 |

* commercial liquid petroleum gas (LPG)

### Liquids

| | Composition by mass % | | | Density at 15°C kg/m$^3$ | Calorific value at 15°C MJ/kg | |
|---|---|---|---|---|---|---|
| | C | H | S | | Gross | Net |
| propane* | 82.0 | 18.0 | | 505 | 50.0 | 46.3 |
| butane* | 81.9 | 17.0 | | 575 | 49.3 | 45.8 |
| petrol (gasoline) | 85.5 | 14.4 | 0.1 | 733 | 46.9 | 43.7 |
| kerosene (parafin) | 85.9 | 14.0 | 0.1 | 780 | 46.5 | 43.4 |
| diesel (gas oil) | 85.7 | 13.4 | 0.9 | 840 | 45.4 | 42.4 |

* commercial liquid petroleum gas (LPG)

# 8. Thermodynamics and Fluid Mechanics

## 8.1 *Temperatures at the primary fixed points*

| | |
|---|---|
| Normal boiling point of oxygen (oxygen point) | -182.97°C |
| Triple point of water | 0.01°C |
| Normal boiling point of water (steam point) | 100.00°C |
| Normal boiling point of sulphur (sulphur point) | 444.60°C |
| Normal melting point of silver (silver point) | 960.80°C |
| Normal melting point of gold (gold point) | 1063.00°C |

## 8.2 *Thermodynamic Relationships*

In all thermodynamic relationships, temperatures and pressures are absolute. Symbols for properties in capitals refer to a given mass of material, those in lower case are per unit mass ("specific").

First Law: $\mathrm{d}Q - \mathrm{d}W = \mathrm{d}U$

Enthalpy: $H = U + pV$ or $h = u + pv$

For a reversible process, $\mathrm{d}S = \left(\frac{\mathrm{d}Q}{T}\right)_{\mathrm{rev}}$ or $\mathrm{d}Q = T\mathrm{d}S$

$\mathrm{d}W = p\mathrm{d}V$

For a homogeneous fluid $T\mathrm{d}s = \mathrm{d}u + p\mathrm{d}v = \mathrm{d}h - v\mathrm{d}p$

Specific heat at constant volume: $c_v = \left(\frac{\mathrm{d}u}{\mathrm{d}T}\right)_v$

Specific heat at constant pressure: $c_p = \left(\frac{\mathrm{d}h}{\mathrm{d}T}\right)_p$

Ratio of specific heats: $\gamma = \frac{c_p}{c_v}$

Helmholtz function: $F = U - TS$ or $f = u - Ts$

Gibbs function, Gibbs free energy: $G = H - TS$ or $g = h - Ts$

**Perfect Gas Relationships**

$pv = RT$ or $p = \rho RT$ $\qquad PV = mRT$

$pv_0 = R_0 T$ $\qquad MR = R_0$

$\Delta U = mc_v(T_2 - T_1)$ $\qquad \Delta H = mc_p(T_2 - T_1)$

$$\Delta S = mc_v \ln\left(\frac{p_2}{p_1}\right) + mc_p \ln\left(\frac{v_2}{v_1}\right)$$

$c_p - c_v = R$ $\qquad \frac{\gamma - 1}{\gamma} = \frac{R}{c_p}$

Continuity equation: $\dot{m} = \rho A c$

Steady Flow Energy equation:

$$\frac{Q-W}{m} = h_2 - h_1 + \tfrac{1}{2}(c_2^2 - c_1^2) + g(z_2 - z_1)$$

Availability (closed system):

$$(A_1 - A_0) = (U_1 + p_0 V_1 - T_0 S_1) - (U_0 + p_0 V_0 - T_0 S_0)$$

Availability (flow process):

$$(B_1 - B_0) = (H_1 - T_0 S_1) - (H_0 - T_0 S_0)$$

Irreversibility: $I = W_{rev} - W$

Van der Waal's equation: $(p + \frac{a}{v^2})(v - b) = RT$

Entropy: $S = k \ln p$, where $k = R_0/N$

Reversible (Carnot) engine efficiency: $1 - \frac{T_{sink}}{T_{source}}$

Engine indicated power: $P_i = p_m v_s N_c$

Maximum work of a reaction:

$$W_{max} = G_{react} - G_{prod} = R_0 T \ln(K_p p^n)$$

For reversible polytropic process in a closed system:

$$p V^n = \text{const}$$

$$W = \frac{(p_1 V_1 - p_2 V_2)}{(n-1)} \qquad (n \neq 1)$$

Additionally, if a perfect gas,

$$W = \frac{mR(T_1 - T_2)}{(n-1)} \qquad \frac{T_2}{T_1} = \left(\frac{p_2}{p_1}\right)^{\frac{n-1}{n}} = \left(\frac{V_1}{V_2}\right)^{n-1}$$

$$Q = \frac{\gamma - n}{\gamma - 1} W$$

Adiabatic reversible (isentropic reversible): $n = \gamma$

Isothermal reversible:

$$n = 1 \qquad W = Q = pV \ln\left(\frac{V_2}{V_1}\right)$$

| Wet Vapours | Air-vapour Mixtures |
|---|---|
| $h = h_f + xh_{fg}$ | Relative humidity (percentage saturation): $\phi = \dfrac{p_s}{p_g}$ |
| $u = u_f + xu_{fg}$ | Specific humidty: |
| $s = s_f + xs_{fg}$ | $w = \dfrac{m_s}{m_a} = 0.622 \dfrac{p_s}{p - p_s}$ |

**Maxwell Relations**

$$\left(\frac{\partial T}{\partial v}\right)_s = -\left(\frac{\partial p}{\partial s}\right)_v \qquad \left(\frac{\partial T}{\partial p}\right)_s = \left(\frac{\partial v}{\partial s}\right)_p$$

$$\left(\frac{\partial p}{\partial T}\right)_v = \left(\frac{\partial s}{\partial v}\right)_T \qquad \left(\frac{\partial v}{\partial T}\right)_p = -\left(\frac{\partial s}{\partial p}\right)_T$$

## 8.3 *Heat Transfer*

Log mean temperature difference: $\Delta T_m = \dfrac{\Delta T_{\text{in}} - \Delta T_{\text{out}}}{\ln\left(\dfrac{\Delta T_{\text{in}}}{\Delta T_{\text{out}}}\right)}$

Heat transfer coefficient: $h = \dfrac{\dot{Q}}{A\Delta T}$

Emissivity: $\varepsilon = \dfrac{q}{q_b}$

*Conduction*:

(one-dimensional): $\dfrac{\dot{Q}}{A} = -k\dfrac{\mathrm{d}T}{\mathrm{d}x}$

(radial): $\dfrac{\dot{Q}}{l} = \dfrac{2\pi k\Delta T}{\ln\left(\dfrac{r_2}{r_1}\right)}$

*Forced convection*: $Nu = f(Re, Pr)$

Turbulent forced convection in a tube: $Nu = 0.023\, Re^{0.8} Pr^{0.4}$

*Free convection*: $Nu = f(Gr, Pr)$

Laminar free convection on a vertical plate: $Nu = 0.59\, (Gr \times Pr)^{0.25}$

*Radiation*:

Stefan-Boltzmann Law: $q_b = \sigma T^4$

Grey body to black or large enclosure: $\dfrac{\dot{Q}}{A} = \sigma\varepsilon_1\left(T_1^4 - T_2^4\right)$

Large parallel grey surfaces: $$\frac{\dot{Q}}{A} = \frac{\sigma(T_1^4 - T_2^4)}{\left(\frac{1}{\varepsilon_1} + \frac{1}{\varepsilon_2} - 1\right)}$$

## 8.4 *Fluid Mechanics*

In fluid mechanics pressures can generally be either absolute or gauge. Either may be used except where specifically noted.

**Statics**

$$\frac{dp}{dz} = -\rho g$$

$$p = \rho g h$$

Normal force $F = \rho g \bar{h}$

$$\varepsilon = \frac{A k^2_{centroid}}{A \bar{y}}$$

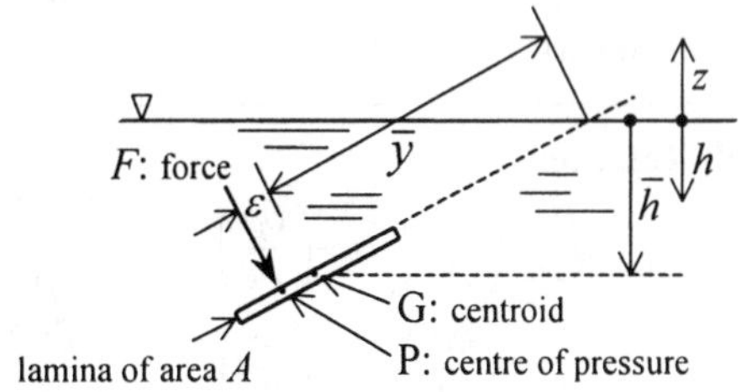


$$\overline{GM}_{roll} = \frac{Ak^2}{V} = \overline{BG}$$

($k^2$ about axis of roll)

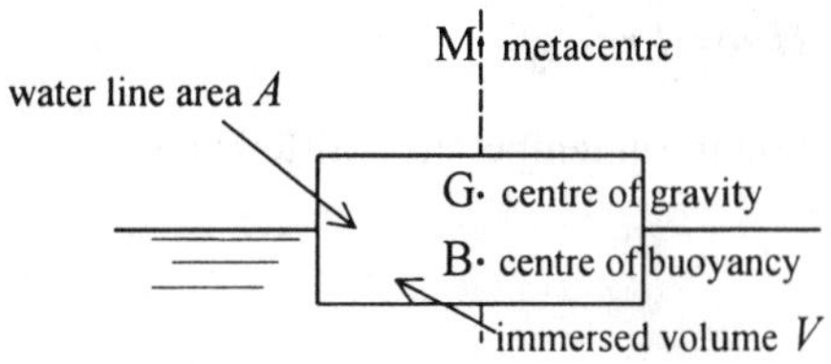


**Dynamics**

Newtonian viscosity: $\tau = \mu \frac{du}{dy}$

Viscous (laminar) pipe flow: $Q = \frac{\pi r^4}{8\mu}\frac{dp}{dx}, \quad f = \frac{16}{Re}$

Stokes' law: $D = 3\pi\mu u d$

Euler's equation: $\frac{1}{\rho}\frac{dp}{ds} + u\frac{du}{ds} + g\frac{dz}{ds} = 0$

Bernoulli's equation: $p + \tfrac{1}{2}\rho u^2 + \rho g h = \text{const} = P$

or $\frac{p}{\rho g} + \frac{u^2}{2g} + z = \text{const} = Z$

Acceleration along a streamline $a_s = u_s \frac{\partial u_s}{\partial s} + \frac{\partial u_s}{\partial t}$

Acceleration normal to a streamline $a_n = \frac{u_s^2}{r} + \frac{\partial u_n}{\partial t}$

Turbulent pipe flow: $\Delta p = 4f \frac{l}{d} \tfrac{1}{2}\rho u^2$

Vortex flow: $\frac{dp}{dr} = \frac{\rho u^2}{r}$

Free vortex: $ur = \text{const}$; Forced vortex: $\frac{u}{r} = \omega = \text{const}$

Constant area flow with friction (Fanno): $\frac{dp}{\rho} + u\,du + 2\frac{fu^2}{d}\,dl = 0$

Reynolds equations for laminar flow in bearings:

$$\frac{\partial}{\partial x}\left(\frac{\rho h^3}{12\mu}\frac{\partial p}{\partial x}\right) + \frac{\partial}{\partial y}\left(\frac{\rho h^3}{12\mu}\frac{\partial p}{\partial y}\right) = \frac{1}{2}\frac{\partial}{\partial x}(\rho u h) + \frac{\partial}{\partial t}(\rho h) + \rho w$$

$$\frac{1}{r}\frac{\partial}{\partial r}\left(\frac{\rho r h^3}{12\mu}\frac{\partial p}{\partial r}\right) + \frac{1}{r^2}\frac{\partial}{\partial \phi}\left(\frac{\rho h^3}{12\mu}\frac{\partial p}{\partial \phi}\right) = \frac{1}{2}\frac{\partial}{\partial \phi}(\rho h \omega) + \frac{\partial}{\partial t}(\rho h) + \rho w$$

Petrov's Law: $T = \frac{\pi h \mu \omega D^3}{4c}$

Turbulent shear stress: $\bar{\tau} = (\mu + \varepsilon)\frac{d\bar{u}}{dy}$

## 8.5 *Boundary Layers:*

**Boundary Layer Thickness**

| Nominal | Displacement | Momentum |
|---|---|---|
| $\delta_{99}$ = value of $y$ at which $u/u_m = 0.99$ | $\delta^* = \int_0^\infty (1 - \frac{u}{u_m})dy$ | $\theta = \int_0^\infty \frac{u}{u_m}(1 - \frac{u}{u_m})dy$ |

Form Factor: $H = \frac{\delta^*}{\theta}$

**Momentum Integral Equation**

$$\tau_0 = \frac{d}{dx}(\rho u_m^2 \theta) + \rho \delta^* u_m \frac{du_m}{dx}$$

If incompressible and zero pressure gradient:

$$c_f = \frac{\tau_0}{\frac{1}{2}\rho u_m^2} = 2\frac{d\theta}{dx}; \quad C_D = \frac{2\theta_L}{L}$$

Transition criterion - typical:

$Re_x = 5 \times 10^5$ ; $Re_\delta = 3500$ ; $Re_\theta = 500$

**Turbulent Boundary Layer** (zero pressure gradient):

"Friction velocity": $$u_\tau = \sqrt{\frac{\tau_0}{\rho}}$$

In the laminar sub-layer: $$\frac{u}{u_\tau} = \frac{y u_\tau}{\nu}$$

The edge of laminar sub-layer is at $$\frac{u}{u_\tau} = \frac{y u_\tau}{\nu} = 11.6 \text{ (approx)}$$

Inner Law (Law of the Wall): $$\frac{u}{u_\tau} = 5.75 \log_{10} \frac{y u_\tau}{\nu} + 5.5$$

Power law approximation: $$\frac{u}{u_\tau} = K\left(\frac{y u_\tau}{\nu}\right)^{\frac{1}{n}} \text{ or } \frac{u}{u_m} = \left(\frac{y}{\delta}\right)^{\frac{1}{n}}$$

Typically $n = 7$, $K = 8.74$, when

$$\frac{\delta^*}{\delta} = \frac{1}{(1+n)} = 0.125 \qquad \frac{\theta}{\delta} = \frac{n}{(1+n)(2+n)} = 0.097$$

$$H = \frac{\delta^*}{\theta} = 1 + \frac{2}{n} = 1.29$$

Blasius' empirical friction formula for smooth surfaces:

$$f = c_f = 0.045 \ Re_x^{-0.25}$$

**Boundary layer growth:**

For zero pressure gradient:

| Laminar: Blasius' exact solution | Turbulent: $\frac{1}{7}$ power law profile and $\frac{1}{4}$ power law friction |
|---|---|
| $\frac{\delta_{99}}{x} = 4.91 \ Re_x^{-0.5}$ | $\frac{\delta}{x} = 0.370 \ Re_x^{-0.2}$ |
| $\frac{\delta^*}{x} = 1.721 \ Re_x^{-0.5}$ | $\frac{\delta^*}{x} = 0.046 \ Re_x^{-0.2}$ |
| $\frac{\theta}{x} = c_f = 0.664 \ Re_x^{-0.5}$ | $\frac{\theta}{x} = 0.036 \ Re_x^{-0.2}$ |
| | $c_f = 0.058 \ Re_x^{-0.2}$ |
| $C_D = 1.328 \ Re^{-0.5}$ | $C_D = 0.072 \ Re^{-0.2}$ |
| $H = \frac{\delta^*}{\theta} = 2.59$ | $H = \frac{\delta^*}{\theta} = 1.29$ |

## 8.6 *Hydraulic machines:*

**Dimensionless coefficients:**

Head: $K_H = \dfrac{gH}{\omega^2 D^2}$ Flow: $K_Q = \dfrac{Q}{\omega D^3}$ Power: $K_P = \dfrac{\text{Power}}{\rho \omega^3 D^5}$

Dimensionless specific speed (rad/s):
$$\omega_s = \left| \frac{\omega Q^{\frac{1}{2}}}{(gH)^{\frac{3}{4}}} = \frac{K_Q^{\frac{1}{2}}}{K_H^{\frac{3}{4}}} = \frac{\omega(\text{power})^{\frac{1}{2}}}{\rho^{\frac{1}{2}}(gH)^{\frac{5}{4}}} \right|_{\eta_{max}}$$

Dimensionless diameter:
$$\Delta = \frac{D(gH)^{\frac{1}{4}}}{Q^{\frac{1}{2}}}$$

NPSE:
$$\frac{p_1 - p_v}{\rho}$$

Dimensionless suction specific speed:
$$K_s = \frac{\omega Q^{\frac{1}{2}}}{(\text{NPSE})^{\frac{3}{4}}}$$

Cavitation number:
$$\sigma \text{ or } k = \frac{p_\infty - p_v}{\frac{1}{2}\rho v_\infty^2}$$

Thoma cavitation number:
$$\sigma_{th} = \frac{p_1 - p_v}{p_2 - p_1} = \left(\frac{\omega_s}{K_s}\right)^{\frac{4}{3}}$$

Suffices in these definitions:
$\infty$ = reference,
$v$ = vapour,
1 = low pressure side,
2 = high pressure side.
(Pressures must be absolute)

**Cordier correlation for hydraulic machines**

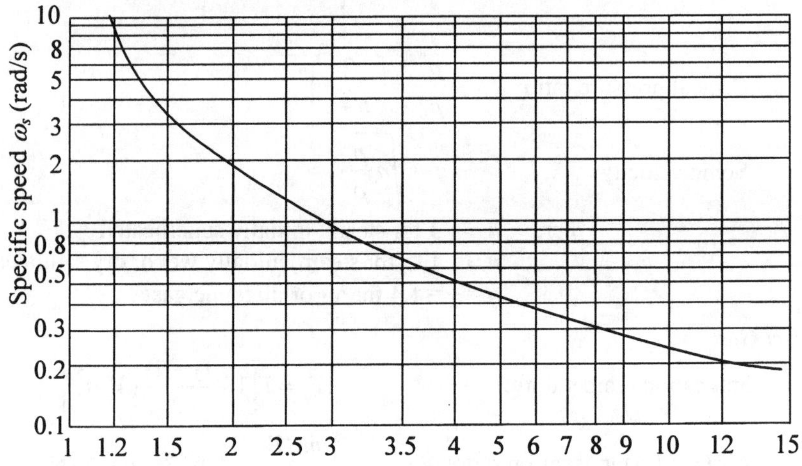

## 8.7 *Open Channel Flows:*

Hydraulic mean depth = hydraulic radius: $m = R = \dfrac{\text{cross-sectional area}}{\text{wetted perimeter}}$

Equivalent diameter: $d' = \dfrac{4 \times \text{cross-sectional area}}{\text{wetted perimeter}} = 4m$

Chézy equation: $u = C\sqrt{RS}$

Manning equation: $u = \dfrac{1}{n} R^{\frac{2}{3}} S^{\frac{1}{2}}$

Steady gradually varied flow: $\dfrac{\mathrm{d}d}{\mathrm{d}x} = \dfrac{S_0 - S_f}{1 - \alpha \dfrac{u^2}{gd}}$ (rectangular channel)

Unsteady gradually varied flow: $\dfrac{\partial d}{\partial x} + \dfrac{u}{g}\dfrac{\partial u}{\partial x} + \dfrac{1}{g}\dfrac{\partial u}{\partial t} = S_0 - S_f$

Continuity: $A\dfrac{\partial u}{\partial x} + u\dfrac{\partial A}{\partial x} + T\dfrac{\partial d}{\partial t} = 0$

(no local inflow or outflow)

Conjugate depths in hydraulic jump: $\dfrac{d_2}{d_1} = \dfrac{1}{2}\left(\sqrt{1 + 8Fr_1^2} - 1\right)$ (rectangular channel)

## 8.8 *High speed gas flow*

*Nozzles*:

Mass flow rate: $\dot{m} = AC_d \sqrt{\dfrac{2n}{(n+1)} p_0 \rho_0 \left[\left(\dfrac{p}{p_0}\right)^{\frac{2}{n}} - \left(\dfrac{p}{p_0}\right)^{\frac{n+1}{n}}\right]}$

Critical pressure ratio: $\dfrac{p^*}{p_0} = \left(\dfrac{2}{n+1}\right)^{\frac{n}{n-1}}$

Sonic velocity: $a = \sqrt{n\dfrac{p}{\rho}}$

where $n \approx 1.3$ for steam, initially superheated
$n \approx 1.135$ for steam, initially wet or dry saturated
$n = \gamma = 1.4$ for air or diatomic gases

*Perfect Gas*:

Stagnation temperature: $T_0 = T\left[1 + \dfrac{(\gamma - 1)}{2}(Ma)^2\right]$

Air ($\gamma = 1.4$) in isentropic flow: $\dfrac{\dot{m}\sqrt{T_0}}{A^* p_0} = 0.0404 \text{ kg K}^{\frac{1}{2}}/(\text{Ns})$

## 8.9 *Dimensionless Groups*

Pressure coefficient: $C_p = \dfrac{p}{\frac{1}{2}\rho u^2}$ or $\dfrac{\Delta p}{\frac{1}{2}\rho u_\infty{}^2}$

Force coefficient (lift, drag etc): $C_F = \dfrac{F}{\frac{1}{2}\rho u^2 A}$, $F = L, D, \ldots$

Discharge coefficient (venturi-meter etc): $C_d = \dfrac{Q_{\text{actual}}}{Q_{\text{ideal}}}$, $Q_{\text{ideal}} = A_{\text{throat}}\sqrt{\dfrac{2\Delta p/\rho}{1-m^2}}$

where $m = A_{\text{throat}}/A_{\text{pipe}}$

Loss coefficient (internal flow) $k = \dfrac{\Delta p}{\frac{1}{2}\rho u^2}$

Pipeflow friction factor (round pipe):
(for non-circular duct or free surface flow, replace $d$ with $d'$ or $4m$) $f = \dfrac{\Delta p}{\frac{1}{2}\rho u^2} \Big/ \left(\dfrac{4l}{d}\right)$

≡ Wall shear stress coefficient: $f = c_f = \dfrac{\tau_0}{\frac{1}{2}\rho u^2}$

| | | | |
|---|---|---|---|
| Fourier number: | $Fo = \dfrac{k}{\rho c_p}\dfrac{t}{l^2}$ | Prandtl number: | $Pr = \dfrac{\mu c_p}{k}$ |
| Froude number: | $Fr = \dfrac{u}{\sqrt{lg}}$ | Rayleigh number: | $Ra = Gr\,Pr$ |
| Grashof number: | $Gr = \dfrac{g\beta\Delta T l^3 \rho^2}{\mu^2}$ | Reynolds number: (general) | $Re = \dfrac{\rho u d}{\mu}$ |
| Knudsen number: | $Kn = \dfrac{\lambda}{l}$ | Reynolds number: (rotating disc) | $Re = \dfrac{\rho\omega D^2}{4\mu}$ |
| Mach number: | $Ma = \dfrac{u}{a}$ | Strouhal number: | $St = \dfrac{nd}{u}$ |
| Nusselt number: | $Nu = \dfrac{hl}{k}$ | Weber number: | $We = u\sqrt{\dfrac{\rho l}{\sigma}}$ |

## 8.10 *Pipe Friction Chart*

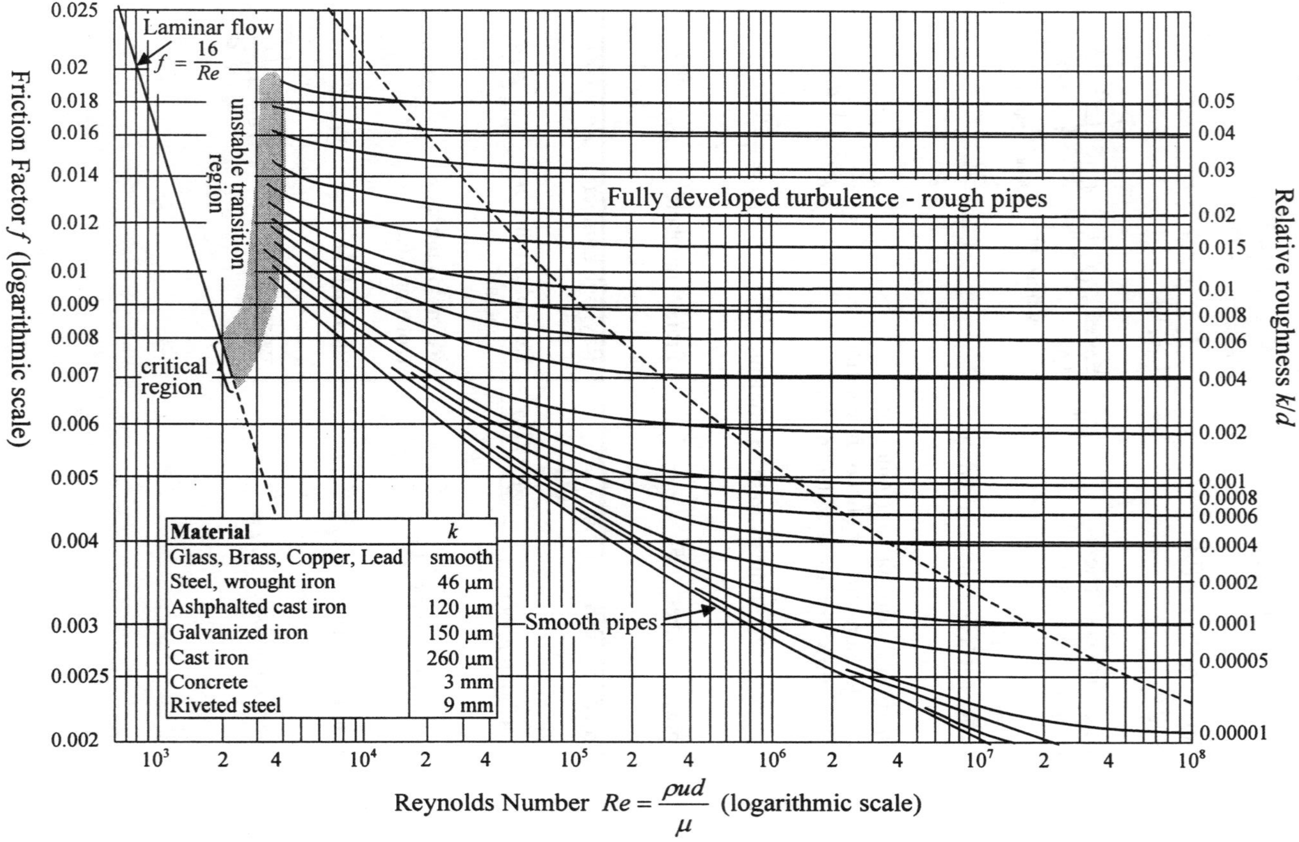
Laminar flow
$f = \frac{16}{Re}$
unstable transition region
critical region
Fully developed turbulence - rough pipes
Smooth pipes
Material | k
Glass, Brass, Copper, Lead | smooth
Steel, wrought iron | 46 μm
Ashphalted cast iron | 120 μm
Galvanized iron | 150 μm
Cast iron | 260 μm
Concrete | 3 mm
Riveted steel | 9 mm
Friction Factor $f$ (logarithmic scale)
0.025
0.02
0.018
0.016
0.014
0.012
0.01
0.009
0.008
0.007
0.006
0.005
0.004
0.003
0.0025
0.002
Relative roughness $k/d$
0.05
0.04
0.03
0.02
0.015
0.01
0.008
0.006
0.004
0.002
0.001
0.0008
0.0006
0.0004
0.0002
0.0001
0.00005
0.00001
$10^3$ 2 4 $10^4$ 2 4 $10^5$ 2 4 $10^6$ 2 4 $10^7$ 2 4 $10^8$
Reynolds Number $Re = \frac{\rho u d}{\mu}$ (logarithmic scale)

# 9. Automatic Control

## 9.1 Block Diagrams

$$\frac{\theta_2}{\theta_1} = G(s) \quad \text{(transfer function)}$$

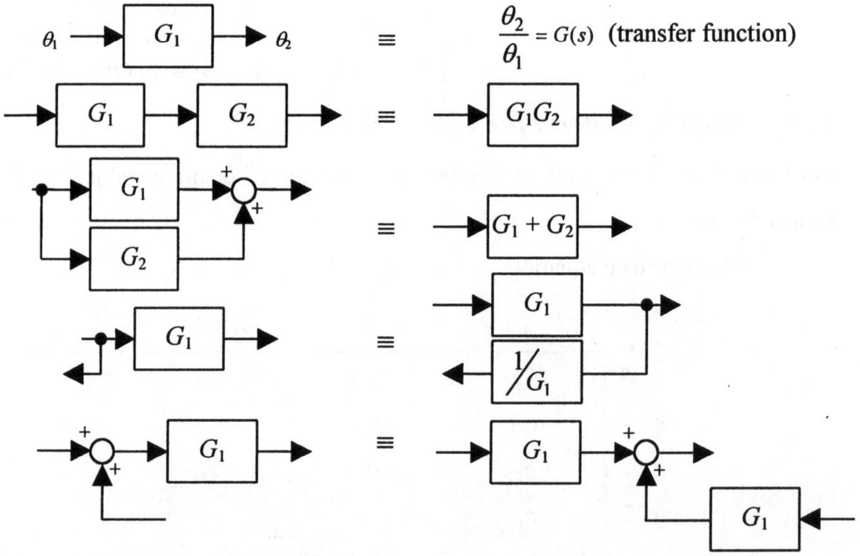

## 9.2 Feedback Notation

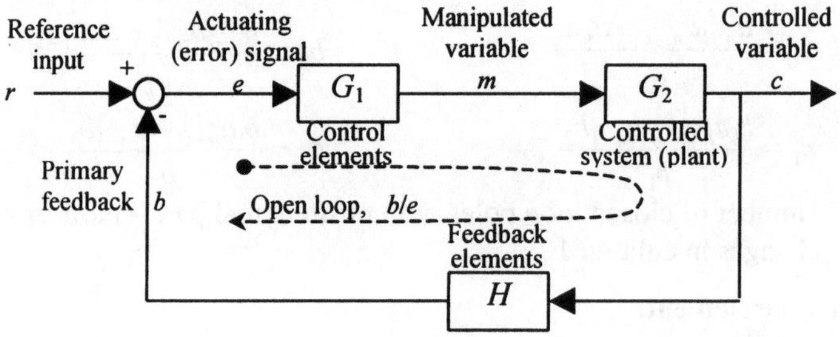

**Transfer functions:** (In the following, $C = L[c]$, etc.)

Input/output closed loop transfer function: $\dfrac{C}{R} = \dfrac{G_1 G_2}{1 + G_1 G_2 H}$

Input/error closed loop transfer function: $\dfrac{E}{R} = \dfrac{1}{1 + G_1 G_2 H}$

Characteristic equation: $1 + G_1 G_2 H = 0$

Open loop transfer function:

$$G_1G_2H \triangleq KGH = \frac{K}{s^\ell}\frac{\prod_{i=1}^{m}(s+z_i)}{\prod_{i=1}^{n-\ell}(s+p_i)}$$

| | |
|---|---|
| $K$ | gain |
| $z_i, p_i$ | zeros and poles |
| $\ell$ | system type number |
| $n$ | system order number |

## 9.3 *Stability Criteria for Linear Systems*

**Root Location:** No closed loop system pole may have a positive real part.

**Routh Array:**

Characteristic equation: $a_n s^n + a_{n-1}s^{n-1} + \dots + a_1 s + a_0 = 0$

| | 1 | 2 | 3 | ... |
|---|---|---|---|---|
| $n$ | $a_n$ | $a_{n-2}$ | $a_{n-4}$ | ... |
| $n-1$ | $a_{n-1}$ | $a_{n-3}$ | $a_{n-5}$ | ... |
| $n-2$ | $b_1$ | $b_2$ | $b_3$ | ... |
| $n-3$ | $c_1$ | $c_2$ | $c_3$ | ... |
| ... | ... | ... | ... | ... |

$$b_1 = \frac{a_{n-1}a_{n-2} - a_n a_{n-3}}{a_{n-1}} \qquad b_2 = \frac{a_{n-1}a_{n-4} - a_n a_{n-5}}{a_{n-1}}$$

$$c_1 = \frac{b_1 a_{n-3} - a_{n-1}b_2}{b_1} \qquad c_2 = \frac{b_1 a_{n-5} - a_{n-1}b_3}{b_1}, \quad \text{etc.}$$

Number of closed loop poles with positive real part = number of sign changes in column 1.

**Nyquist encirclement:**

$P = N + Z$

$N$ = number of clockwise encirclements of (-1, j0) by open loop locus
$P$ = number of closed loop poles with positive real part
$Z$ = number of open loop poles with positive real part

**Gain Margin:** $|KG(j\omega_g)H(j\omega_g)|^{-1}$, for $\omega_g$ such that

$$\angle KG(j\omega_g)H(j\omega_g) = -180°$$

**Phase Margin:** $\angle KG(j\omega_p)H(j\omega_p) + 180°$, for $\omega_p$ such that

$$|KG(j\omega_p)H(j\omega_p)| = 1$$

## 9.4 *Rules of Root Locus sketching*

1. Every point $\alpha$ on the root locus for positive $K$ satisfies

$$|G(\alpha)H(\alpha)| = 1/K$$

$$\angle G(\alpha)K(\alpha) = (1 + 2k)\times 180^\circ, \qquad k = 0, \pm 1, \pm 2, \ldots$$

2. The number of branches of the root locus is equal to the number of poles.
3. Branches of the root locus can be considered to start on the poles ($K = 0$) and terminate on zeros ($K = \infty$) of the open loop system.
4. Points of the root locus exist on the real axis to the left of an odd number of poles plus zeros.
5. The locus is symmetrical with respect to the real axis.
6. The angles of asymptotes $\alpha_k$ to the root locus are given by

$$\alpha_k = \pm\frac{(2k+1)}{n-m}\pi, \qquad k = 0, 1, 2, \ldots$$

($n$ = number of poles, $m$ = number of zeros)

7. The intersection of the asymptotes and the real axis occurs at

$$s_r = -\frac{\sum p_i - \sum z_i}{n-m}$$

8. The locus leaves the real axis or arrives on it at points $\alpha$, given by

$$\frac{\mathrm{d}}{\mathrm{d}\alpha}[KG(\alpha)H(\alpha)] = 0$$

9. The intersection of the root locus and the imaginary axis can be found by application of Routh's stability criteria.

## 9.5 Table of Characteristic Systems

| Typical Example: Electrical | Typical Example: Dynamic | Basic form of transfer function $G(s)$ | Step Response | Frequency Response: Complex Plane (Nyquist) | Frequency Response: Logarithmic (Bode) | Pole-Zero map and Root Locus |
|---|---|---|---|---|---|---|
| | | $K$ | | | | Not applicable |
| | | $\frac{1}{s}$ | | | | |
| | | $\frac{1}{1+Ts}$ | | | | |
| | | $\frac{1}{s(1+Ts)}$ | | | | |
| | | $\frac{\omega_n^2}{s^2+2\xi\omega_n s+\omega_n^2}$ | | | | |
| | | $\frac{1+\alpha Ts}{1+Ts}$ | | | | |
| | | $\frac{\omega_n^2(1+2\xi s/\omega_n)}{s^2+2\xi\omega_n s+\omega_n^2}$ | | | | |

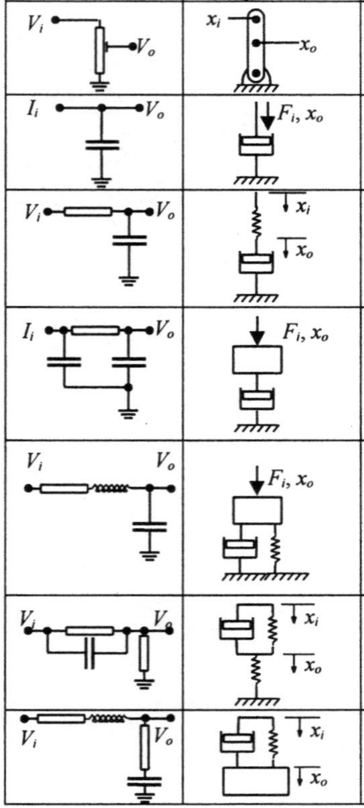


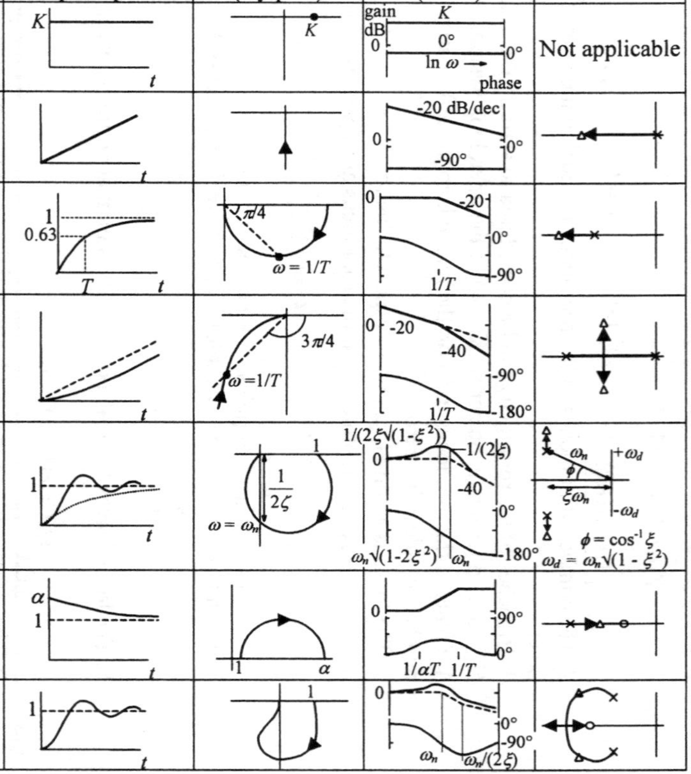

# 10. Electricity and Magnetism

**Ohm's Law**

$V = IR \qquad R = \frac{V}{I} \qquad I = \frac{V}{R}$

**Power**

DC Power $= VI = I^2R = \frac{V^2}{R}$

AC Power $= \mathrm{Re}(\underline{V} \bullet \underline{I}^*) = |V||I|\cos\phi$

**Resistance**

$I = \frac{a}{\rho_0(1+\alpha(T-T_0))}\frac{\mathrm{d}V}{\mathrm{d}x} \qquad R = \int \frac{\rho_0(1+\alpha(T-T_o))}{a}\mathrm{d}x$

**Inductance**

$E = -L\frac{\mathrm{d}I}{\mathrm{d}t} \qquad I = -\int \frac{V}{L}\mathrm{d}t$

$L = N^2\mu_0\mu_r\frac{a}{\ell} \qquad$ Stored energy $= \tfrac{1}{2}LI^2$

L R circuit decay: $I = I_0 e^{-Rt/L}$

**Capacitance**

$Q = CV = \int I\mathrm{d}t \qquad I = \frac{\mathrm{d}Q}{\mathrm{d}t} = C\frac{\mathrm{d}V}{\mathrm{d}t}$

$C = \varepsilon_0\varepsilon_r(n-1)\frac{a}{d}$ for $n$ parallel plates

Stored energy $= \tfrac{1}{2}CV^2 \qquad F = \tfrac{1}{2}\varepsilon_0\varepsilon_r a\left(\frac{V}{x}\right)^2$

**Electrostatics**

$F = \frac{Q_1Q_2}{4\pi\varepsilon_0 r^2} \qquad \underline{F} = e\underline{E} = -e\,\mathrm{grad}\,V$

$Q = \oint \underline{D} \bullet d\underline{S} \quad (=\Psi) \quad \underline{D} = \varepsilon_0\varepsilon_r\underline{E}$

**Electromagnetism**

$E = -N\frac{\mathrm{d}\Phi}{\mathrm{d}t} \qquad B = \frac{\mu_0 I}{2\pi r}$

$F = B\ell I$ (perpendicular components only) $\qquad F = \frac{\mu_0 I_1 I_2 \ell}{2\pi d}$

$\frac{\mathrm{d}H}{\mathrm{d}\ell} = \frac{I\sin\alpha}{4\pi x^2} \qquad$ for solenoid: $H = \frac{NI}{\ell}$

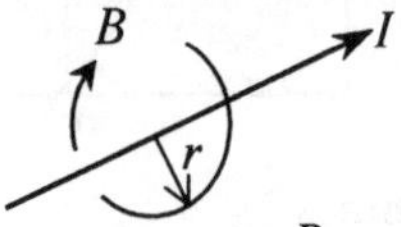


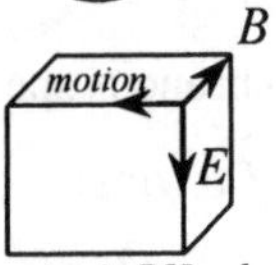

*Fleming's RH rule*

**Magnetism**

$$H = \frac{B}{\mu_0 \mu_r}$$

For a magnetic circuit, $B = \frac{\Phi}{a}$ $\Phi = \frac{NI}{\frac{\ell_1}{\mu_1 a_1} + \frac{\ell_2}{\mu_2 a_2}}$

Stored energy density $= \frac{1}{2} HB = \frac{1}{2}\frac{B^2}{\mu_0}$, $F = \left(\frac{1}{2} HB\right) a = \frac{B^2 a}{2\mu_0}$

**DC/Universal machine characteristics**:

$E \propto n\Phi$ $T \propto I_a \Phi$ $V = E \pm I_a R_a$

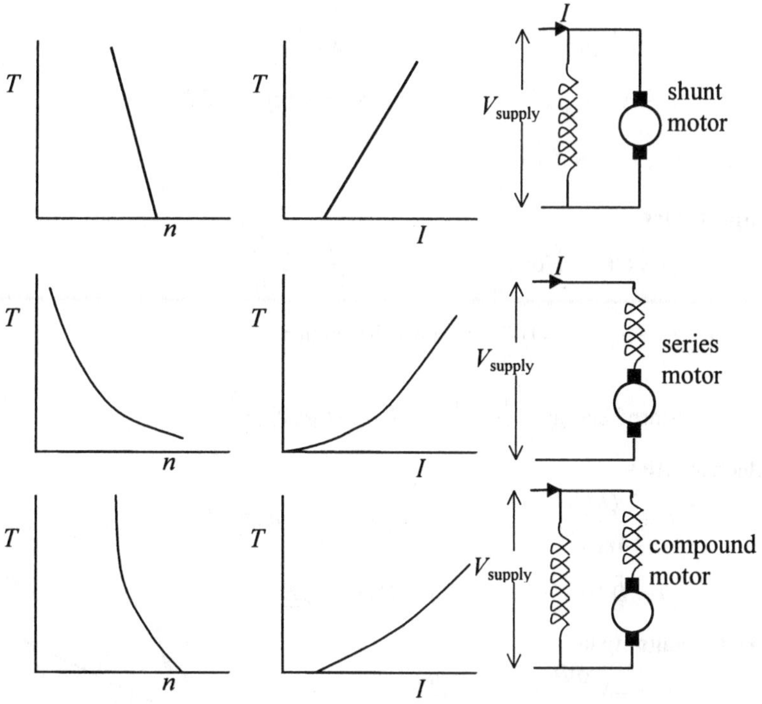


**AC Machines**

Synchronous speed $= f/p$

$E \propto f\Phi$ (rms)

$$T \propto \frac{\Phi^2 sR}{R^2 + (s X_0)^2}$$

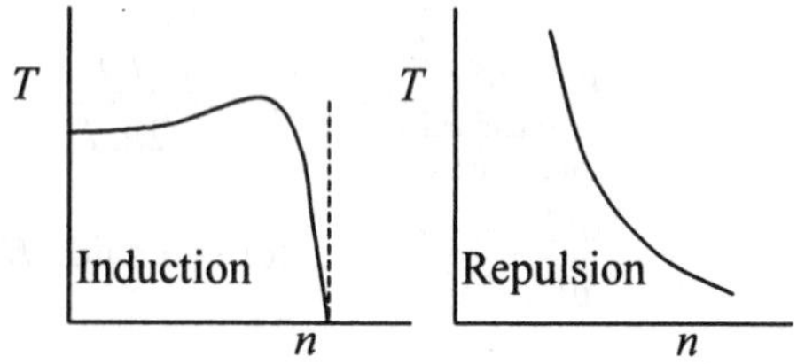

## AC Circuits

$$V_{rms} = \frac{1}{\sqrt{2}} V_{max} \qquad \omega = 2\pi f$$

Series LCR:

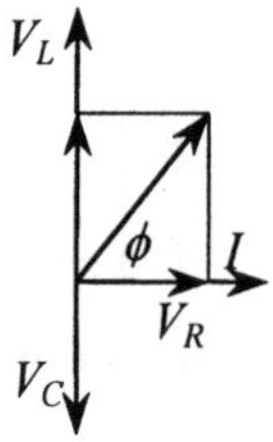

$$Z = \left[ R^2 + \left( \omega L - \frac{1}{\omega C} \right)^2 \right]^{0.5}$$

$$\underline{Z} = R + j\omega L + \frac{1}{j\omega C} = R + j\left( \omega L - \frac{1}{\omega C} \right) \qquad \cos\phi = \frac{R}{Z}$$

At resonance, $\quad \omega = \omega_o = \dfrac{1}{\sqrt{LC}} \qquad\qquad Q \text{ factor} = \omega_o \dfrac{L}{R}$

## Basic 'Op Amp' Circuits

$$V_o \approx -\frac{R_2}{R_1} V_i$$

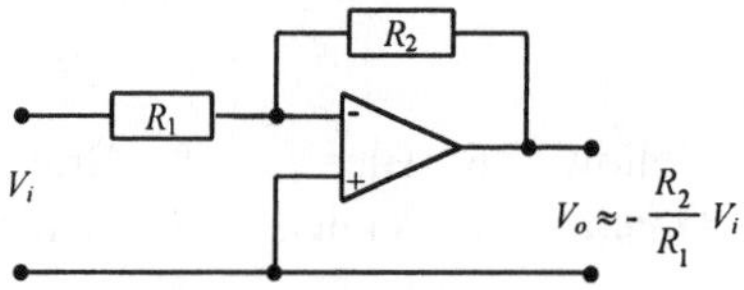

Inverting Amplifier

$$V_o \approx (1 + \frac{R_2}{R_1}) V_i$$

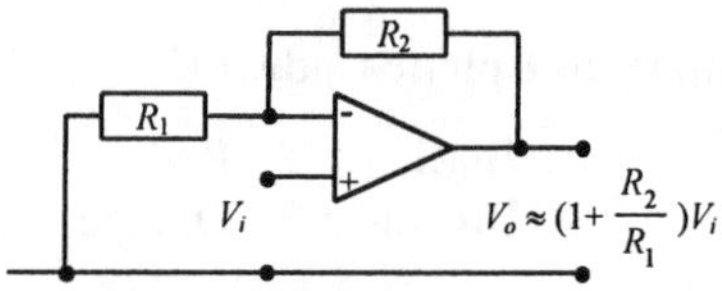

Non-inverting amplifier

$$V_o \approx \frac{R_2}{R_1}(V_2 - V_1)$$
$$\text{with } \frac{R_1}{R_2} = \frac{R_3}{R_4}$$

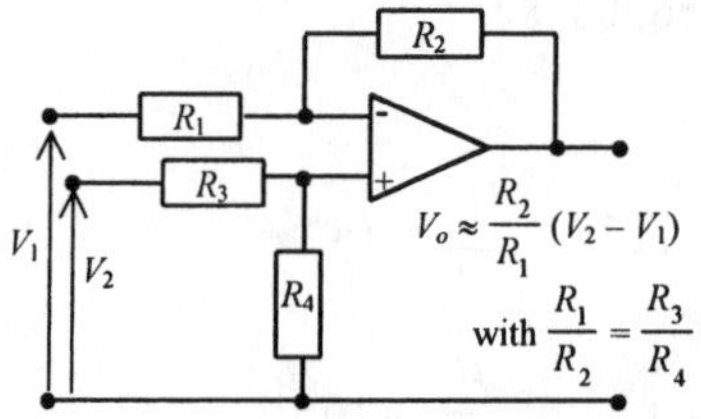

Differential Input Amplifier

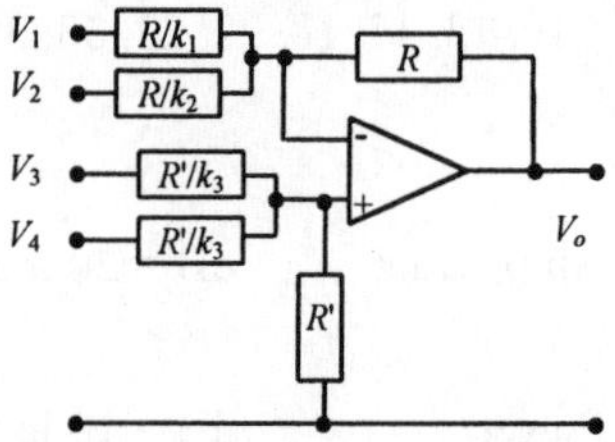

$$V_o = (k_3 V_3 + k_4 V_4) - (k_1 V_1 + k_2 V_2)$$

Adding/Subtracting/Scaling

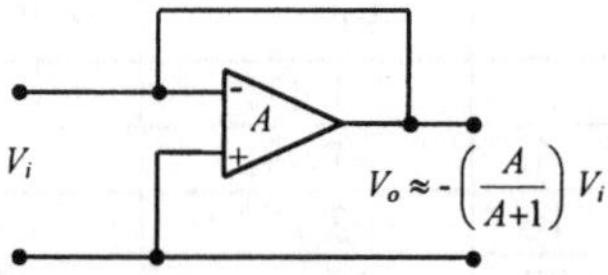

$$V_o \approx -\left( \frac{A}{A+1} \right) V_i$$

Voltage follower

$$V_o = IR$$

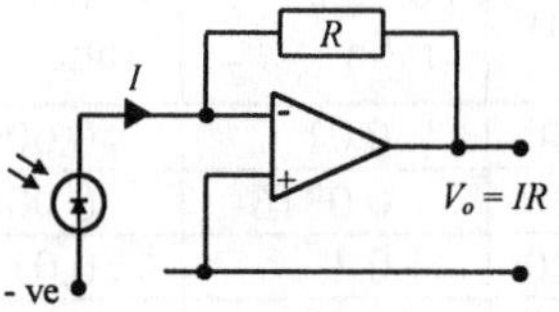

Current amplifier

$$V_o \approx -\frac{1}{RC} \int_0^t V_i \, dt$$

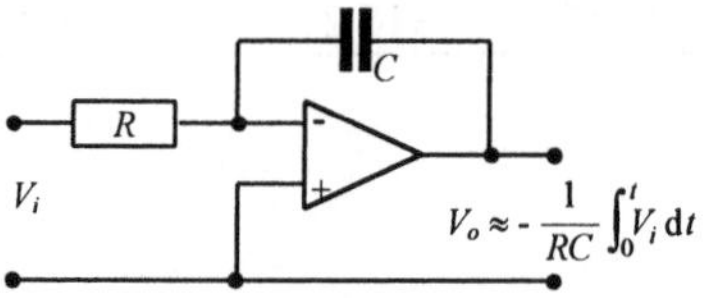

Integrating amplifier

$$V_o \approx -RC \frac{dV_i}{dt}$$

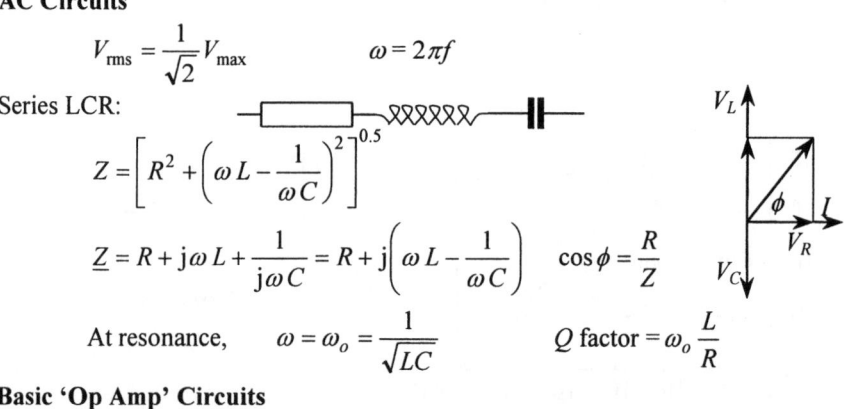

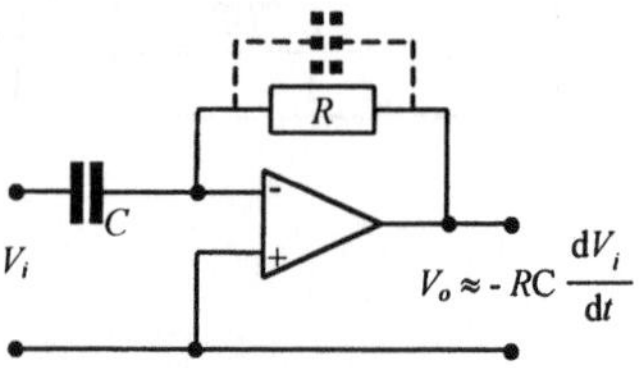

Differentiating amplifier

**μA741 Op Amp Characteristics**

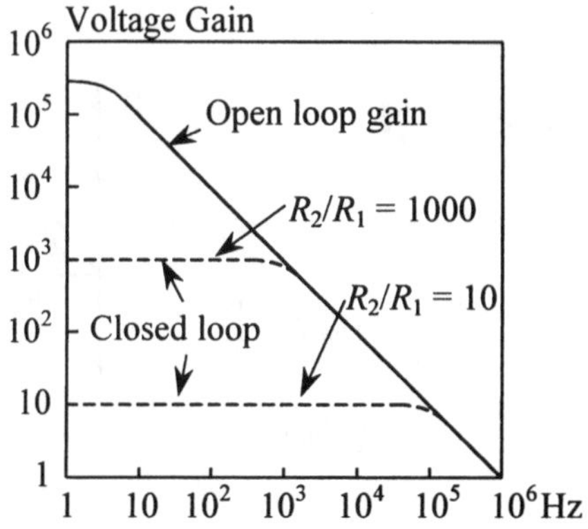


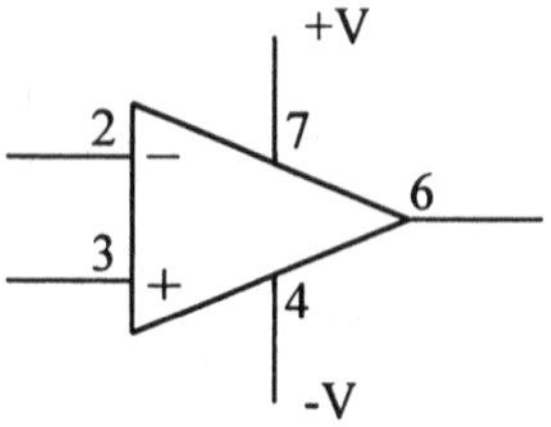


**Component Colour Code**

| | | | | | | | | | |
|---|---|---|---|---|---|---|---|---|---|
| 0 | Black | 2 | Red | 4 | Yellow | 6 | Blue | 8 | Grey |
| 1 | Brown | 3 | Orange | 5 | Green | 7 | Violet | 9 | White |

**Component Preferred Values**

10, 12, 15, 18, 22, 27, 33, 39, 47, 56, 68, 82

**Decibels**

Amplitude: $\mathrm{dB} = 20 \log_{10}\left(\frac{V_1}{V_2}\right)$ $\qquad \frac{V_1}{V_2} = 10^{\mathrm{dB}/20}$

Power: $\mathrm{dB} = 10 \log_{10}\left(\frac{W_1}{W_2}\right)$ $\qquad \frac{W_1}{W_2} = 10^{\mathrm{dB}/10}$

| dB | Voltage ratio $\frac{V_1}{V_2}$ | Power ratio $\frac{W_1}{W_2}$ | dB | Voltage ratio $\frac{V_1}{V_2}$ | Power ratio $\frac{W_1}{W_2}$ |
|---|---|---|---|---|---|
| -40 | 0.01 | 0.0001 | 3 | 1.412 | 1.995 |
| -30 | 0.0316 | 0.001 | 6 | 1.995 | 3.981 |
| -20 | 0.1 | 0.01 | 10 | 3.162 | 10 |
| -10 | 0.316 | 0.1 | 20 | 10 | 100 |
| -6 | 0.501 | 0.251 | 30 | 31.62 | 1000 |
| -3 | 0.708 | 0.501 | 40 | 100 | 10 000 |
| 0 | 1 | 1 | 50 | 316.2 | 100 000 |

# 11 Soil Mechanics

## 11.1 *Soil Classification*

**Particle size**

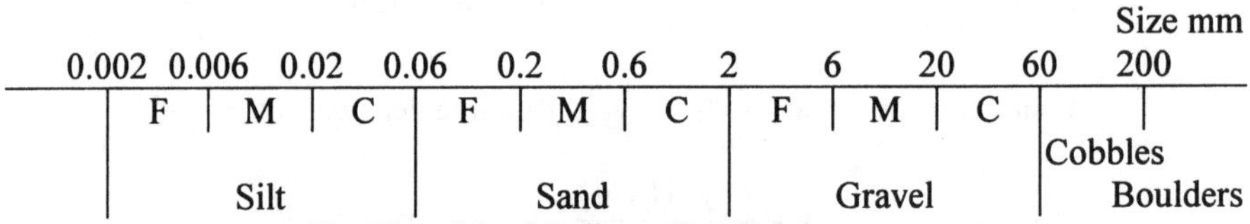


F = Fine, M = Medium, C = Coarse

**Casagrande Soil Classification** for fine grained soils

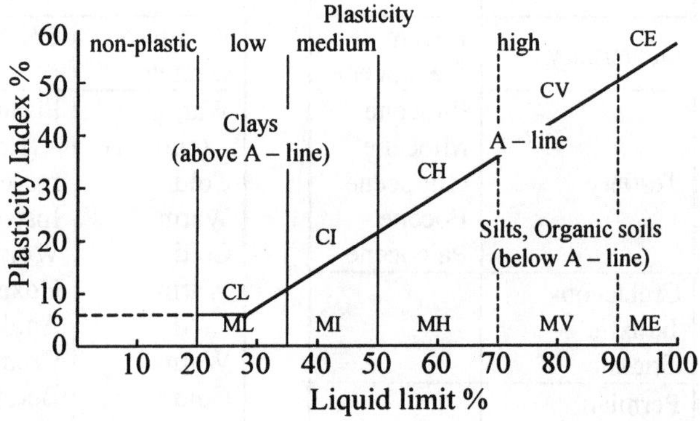


| | | |
|---|---|---|
| C = Clay | L = Low plasticity | $W_{LL} < 35\%$ |
| M = Silt | I = Intermediate (medium) plasticity | $35\% \leq W_{LL} < 50\%$ |
| O = Organic | H = High plasticity | $50\% \leq W_{LL} < 70\%$ |
| Pt = Peat | V = Very high plasticity | $70\% \leq W_{LL} < 90\%$ |
| | E = Extremely high plasticity | $W_{LL} \geq 90\%$ |

## 11.2 *Volume – weight relationships*

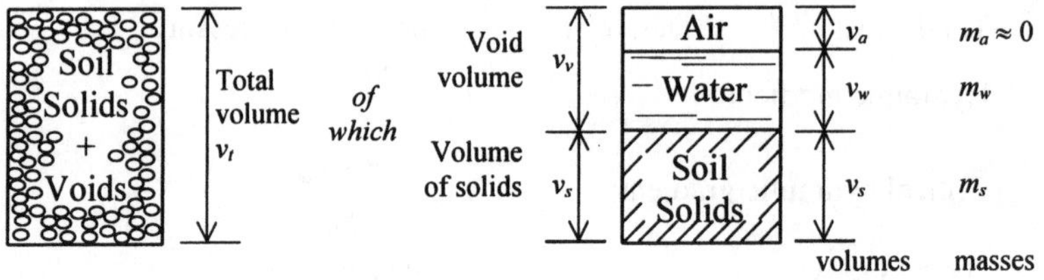

**Definitions:**

| | | | |
|---|---|---|---|
| Void ratio | $e = \frac{v_v}{v_s}$ | Specific volume | $v = \frac{v_t}{v_s} = 1 + e$ |
| Porosity | $n = \frac{v_v}{v_t} = \frac{e}{1+e}$ | Saturation ratio | $S_r = \frac{v_w}{v_v}$ |
| Water content | $w = \frac{m_w}{m_s}$ | Relative density | $D_r = \frac{e_{\max} - e}{e_{\max} - e_{\min}}$ |
| $wG_s = eS_r$ | $\gamma = \frac{G_s \gamma_w (1+w)}{v}$ | | $\gamma_{dry} = \frac{G_s \gamma_w}{v}$ |

## 11.3 *Stratigraphic Table*

| Era | Period | Epoch |
|---|---|---|
| Cenozoic | Quaternary | Recent<br>Pleistocene |
| | Tertiary | Pliocene<br>Miocene<br>Oligocene<br>Eocene<br>Paleocene |
| Mesozoic | Cretaceous<br>Jurassic<br>Triassic | |
| Paleozoic | Permian<br>Carboniferous<br>Devonian<br>Silurian<br>Ordovician<br>Cambrian | |
| Precambrian | | |

| Subdivisions of Quaternary | |
|---|---|
| Relative Climate | UK Name |
| Warm (current) | Flandrian (Holocene) |
| Cold | Devensian |
| Warm | Ipswichian |
| Cold | Wolstonian |
| Warm | Hoxnian |
| Cold | Anglian |
| Warm | Cromerian |
| Cold | Beestonian |
| Warm | Pastonian |
| Cold | Baventian |
| Warm | Antian |
| Cold | Thurnian |
| | Ludhamian |
| | Waltonian |

## 11.4 *Effective Stress and Seepage*

$$\sigma = \sigma' + u$$

Total stress = Effective stress + pore water pressure

Hydraulic gradient: $i = -\frac{dh}{dx}$

Critical hydraulic gradient: $i_{crit} = \frac{\gamma - \gamma_w}{\gamma_w}$

Darcy's Law: $q = Aki$ or $v = ki$

Flownet analysis: $q = kH\dfrac{N_f}{N_h}$ per metre length or thickness

Transformation factor in anisotropic soils: $x' = \sqrt{\dfrac{k_y}{k_x}}$

Transformed permeability: $k_t = \sqrt{k_x k_y}$

Deflection of flowlines at interface: $\dfrac{\tan\beta_1}{\tan\beta_2} = \dfrac{k_1}{k_2}$

## 11.5 *Stress Parameters*

Average effective stress: $p' = \frac{1}{3}(\sigma_1' + \sigma_2' + \sigma_3')$

Deviator stress:

$$q = \frac{1}{\sqrt{2}}\left\{(\sigma_1' - \sigma_2')^2 + (\sigma_1' - \sigma_3')^2 + (\sigma_2' - \sigma_3')^2\right\}^{\frac{1}{2}}$$

$$= (\sigma_1' - \sigma_3') \text{ if } \sigma_1' = \sigma_2' \text{ or } \sigma_2' = \sigma_3' \quad \text{(triaxial conditions)}$$

Stress ratio: $\eta = \dfrac{q}{p'}$

Plane strain: $s' = \frac{1}{2}(\sigma'_1 + \sigma'_3)$ $t = \frac{1}{2}(\sigma'_1 - \sigma'_3)$

## 11.6 *Failure Criteria*

In terms of effective stresses: $\left(\dfrac{\tau}{\sigma'}\right)_{\max} = \tan\phi'$

In terms of total stresses: $\tau_{\max} = \tau_u$ (undrained shear strength)

## 11.7 *Consolidation-time curves*

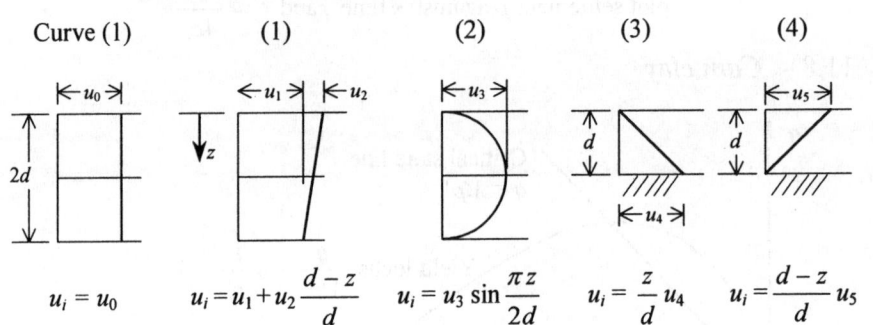

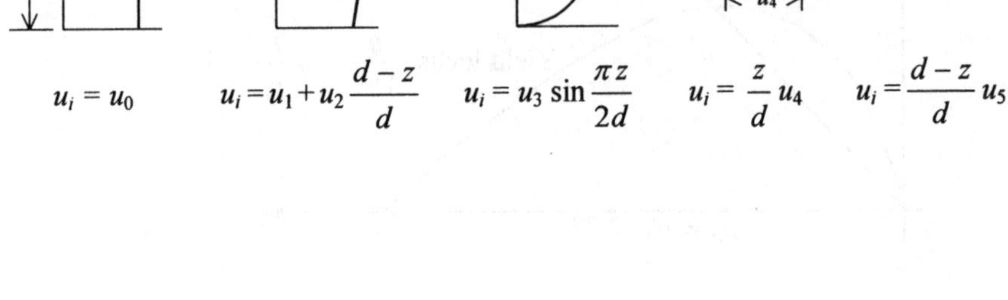


$u_i = u_0$ $u_i = u_1 + u_2\dfrac{d - z}{d}$ $u_i = u_3 \sin\dfrac{\pi z}{2d}$ $u_i = \dfrac{z}{d}u_4$ $u_i = \dfrac{d - z}{d}u_5$

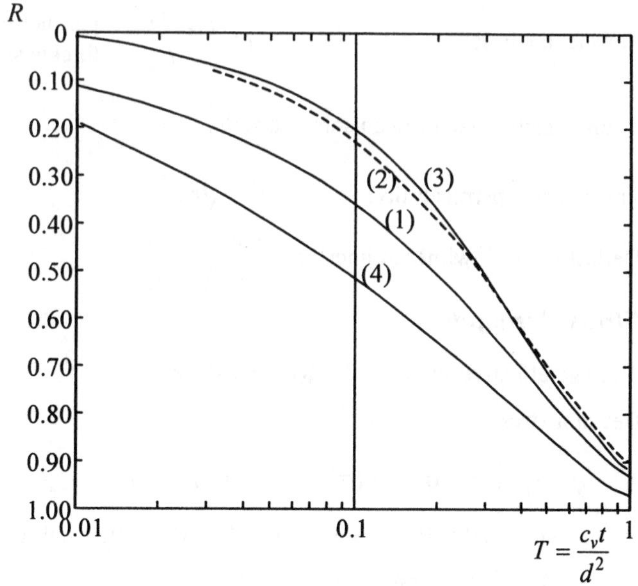


For curve (1): $T = \frac{\pi}{4} R^2$ $\quad R < 0.60$

$T = -0.933 \log_{10}(1 - R) - 0.085 \quad R > 0.60$

$T_{50} = 0.197; \quad T_{90} = 0.848$

Consolidation coefficient: $c_v = \frac{kE'_0}{\gamma_w}$

$t_x$ method for calculating $c_v$ from oedometer test data:

plot settlement $\rho$ against $\sqrt{\text{time}}$, and $t_x = \frac{3d^2}{4c_v}$

## 11.8 *Cam clay*

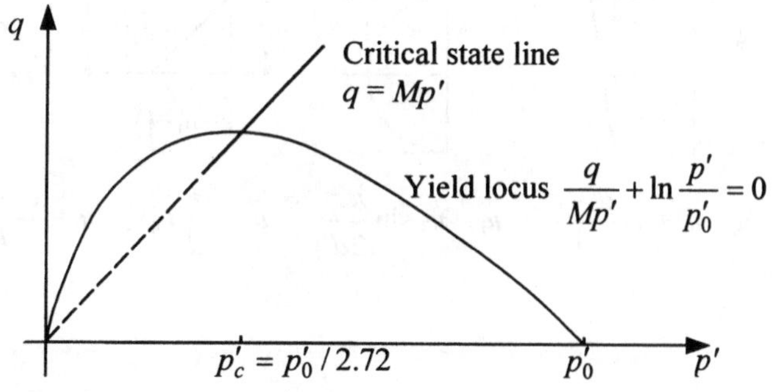

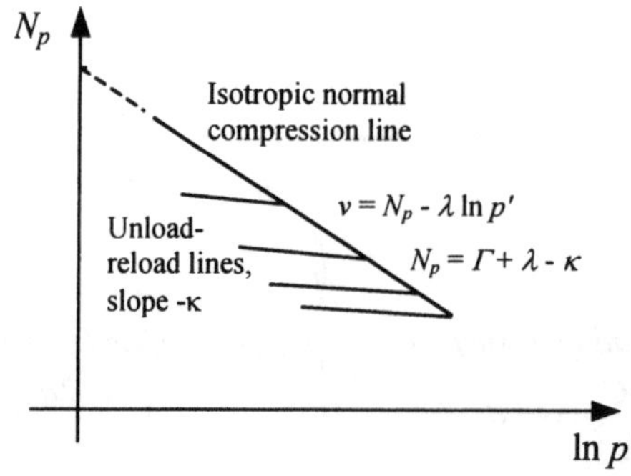


## 11.9 *Stresses and Displacements in Elastic Half-space*

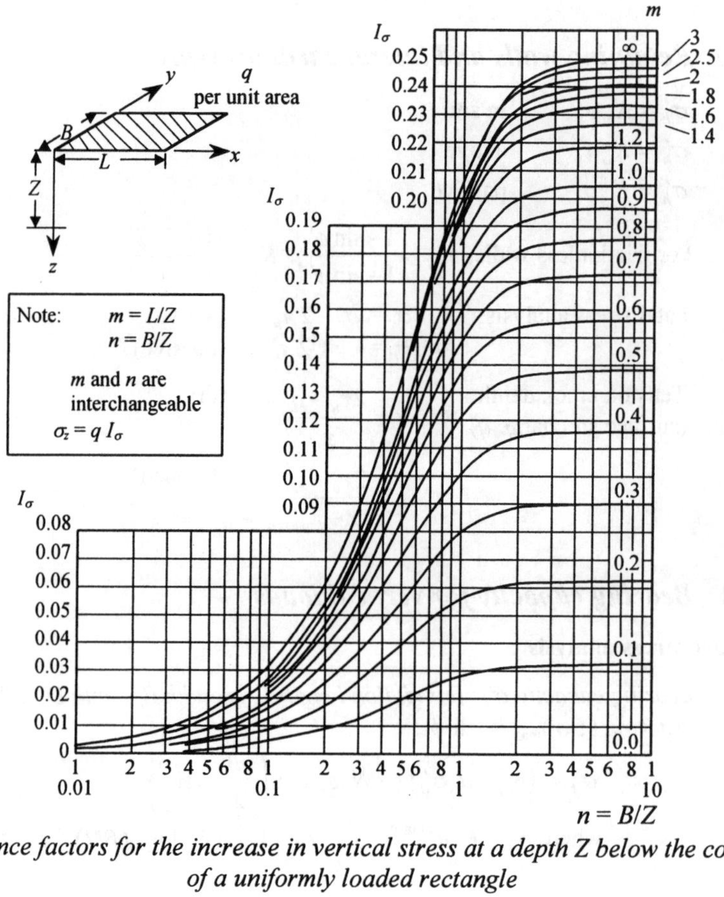


*Influence factors for the increase in vertical stress at a depth Z below the corner of a uniformly loaded rectangle*

**Boussinesq**

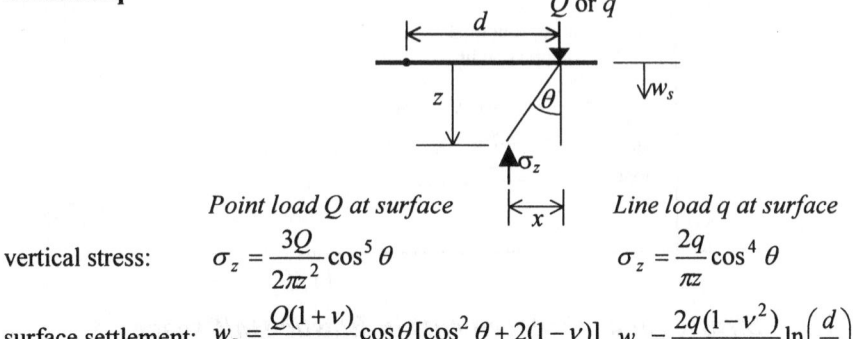

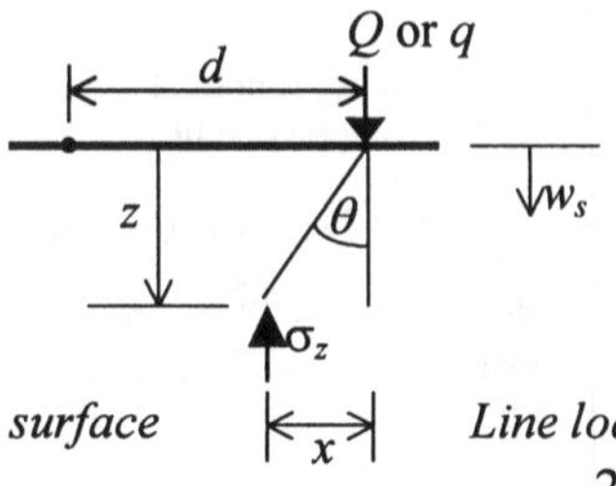


| | *Point load Q at surface* | *Line load q at surface* |
|---|---|---|
| vertical stress: | $\sigma_z = \frac{3Q}{2\pi z^2}\cos^5\theta$ | $\sigma_z = \frac{2q}{\pi z}\cos^4\theta$ |
| surface settlement: | $w_s = \frac{Q(1+\nu)}{2\pi z E}\cos\theta[\cos^2\theta + 2(1-\nu)]$ | $w_s = \frac{2q(1-\nu^2)}{\pi E}\ln\left(\frac{d}{x}\right)$ |

where displacement at $d$ is assumed = 0 ($d \geq x$).

## 11.10 *Retaining walls and lateral earth pressure*

$\sigma'_h = K_o\sigma'_v$ in situ

$\sigma'_h = K_a\sigma'_v$ active

$\sigma'_h = K_p\sigma'_v$ passive

For frictionless walls, $K_a = \frac{1-\sin\phi'}{1+\sin\phi'}$; $K_p = \frac{1+\sin\phi'}{1-\sin\phi'}$

Total stress analysis: $\sigma_h = \sigma_v - 2\,\tau_u$ (active)

$\sigma_h = \sigma_v + 2\,\tau_u$ (passive)

Tension crack depth: (surface surcharge $q$)

$z = \frac{2\tau_u - q}{\gamma}$ (dry)

$z = \frac{2\tau_u - q}{\gamma - \gamma_w}$ (flooded)

## 11.11 *Bearing capacity for vertical loads*

**Effective stress analysis**

*Bearing capacity* $\sigma'_f$*: foundation length L, breadth B, depth D, failure criterion* $(\tau/\sigma')_{max} = \tan\phi'$ :

$$\sigma'_f = \{N_q s_q d_q \sigma'_o\} + \left\{N_\gamma s_\gamma d_\gamma r_\gamma\left(\frac{\gamma B}{2} - \Delta u\right)\right\}$$

where $N_q = K_p e^{\pi\tan\phi'}$ $\quad r_\gamma = 1 - 0.25\log_{10}(B/2)$ for $B \geq 2$ m

| Parameter | Meyerhof ($\phi' > 10°$) | Brinch Hansen |
|---|---|---|
| Shape factor $s_q$ | $1 + 0.1\,K_p\,(B/L)$ | $1 + [(B/L)\tan\phi']$ |
| Depth factor $d_q$ | $1 + 0.1\,\sqrt{K_p}\,(D/B)$ | $1 + 2\tan\phi'(1 - \sin\phi')k$ |
| | | $k = D/B$    $D/B \le 1$ |
| | | $k = \tan^{-1}(D/B)$ (radians)    $D/B > 1$ |
| $N_\gamma$ | $(N_q - 1)\tan(1.4\phi')$ | $1.5\,(N_q - 1)\tan\phi'$ |
| Shape factor $s_\gamma$ (for $N_\gamma$) | $= s_q$ | $1 - 0.4\,(B/L)$ |
| Depth factor $d_\gamma$ (for $N_\gamma$) | $= d_q$ | 1 |

**Total stress analysis**

*Bearing capacity $\sigma_f$ : foundation length L, breadth B, depth D, failure criterion* $\tau_{\max} = \tau_u$ :

$$(\sigma_f - \sigma_o) = N_c s_c d_c \tau_u \quad \text{where} \quad N_c = 5.14$$

| Parameter | Skempton | Meyerhof |
|---|---|---|
| Shape factor $s_c$ | $1 + 0.2\,(B/L)$ | $1 + 0.2\,(B/L)$ |
| Depth factor $d_c$ | $1 + 0.23\,\sqrt{(D/B)}$ up to a maximum of 1.46 ($D/B = 4$) | $1 + 0.2\,(D/B)$ |

## 11.12 *Slope Stability*

Pore pressure ratio: $r_u = \dfrac{u}{\gamma h}$

$$\tau_{mob} = \frac{\tau_u}{F} \qquad \text{(total stress analysis)}$$

$$\tan\phi'_{mob} = \frac{\tan\phi'}{F} \qquad \text{(effective stress analysis)}$$

($F$ = factor of safety)

Circular slips: $$F = \frac{R\sum(\tau_u l)}{\sum x_w w} \qquad \text{(total stress analysis)}$$

$$F = \frac{\sum[(w\cos\alpha - ul)\tan\phi']}{\sum w\sin\alpha} \qquad \text{(effective stress analysis)}$$

Bishop's simplified method: $$F = \frac{1}{\sum w\sin\alpha}\sum\left[\frac{(w - ub)\tan\phi'}{\cos\alpha + \dfrac{\tan\phi'\sin\alpha}{F}}\right]$$

# 12. Structures

The section second moments of area required in many formulae in this section may be found from the formulae on pages 5-1 to 5-2.

## 12.1 *Elastic bending and torsion relationships*

For elastic, isotropic materials,

$$G = \frac{E}{2(1+\nu)} \qquad K = \frac{E}{3(1-2\nu)}$$

Simple bending: $$\frac{M}{I} = \frac{\sigma}{y} = \frac{E}{R} \qquad \sigma = \frac{M}{Z} \qquad Z = \frac{I}{h_{max}}$$

$$M = -EI\frac{d^2y}{dx^2} \qquad Q = \frac{dM}{dx}$$

Shear stress in beam: $$\tau = \frac{QA\bar{y}}{bI}$$

Torsion of a circular section: $$\frac{T}{J} = \frac{\tau}{r} = \frac{G\theta}{L}$$

For a solid section, $$J = \frac{\pi d^4}{32} = A\frac{d^2}{8}$$

For a tube, $$J = \frac{\pi}{32}(d_2^4 - d_1^4) = \frac{A}{8}(d_2^2 + d_1^2)$$

($d_1$ = inner diameter, $d_2$ = outer diameter)

## 12.2 *Buckling Loads*

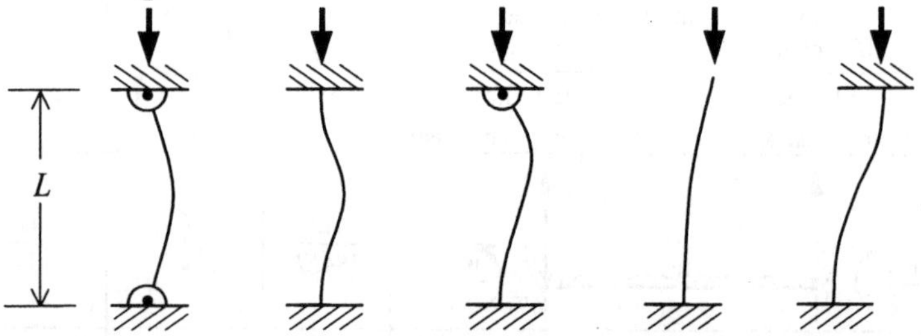

| | | | | | |
|---|---|---|---|---|---|
| Buckling Load | $\dfrac{\pi^2 EI}{L^2}$ | $\dfrac{4\pi^2 EI}{L^2}$ | $\dfrac{2.045\pi^2 EI}{L^2}$ | $\dfrac{\pi^2 EI}{4L^2}$ | $\dfrac{\pi^2 EI}{L^2}$ |
| Effective Length | $L$ | $0.5L$ | $0.699L$ | $2L$ | $L$ |

## 12.3 *Beams bent about principal axis*

| $L$ = overall length<br>$W$ = point load, $M$ = moment<br>$w$ = load per unit length | End Slope | Max Deflection | Max bending moment |
|---|---|---|---|
|  | $\dfrac{ML}{EI}$ | $\dfrac{ML^2}{2EI}$ | $M$ |
|  | $\dfrac{WL^2}{2EI}$ | $\dfrac{WL^3}{3EI}$ | $WL$ |
| 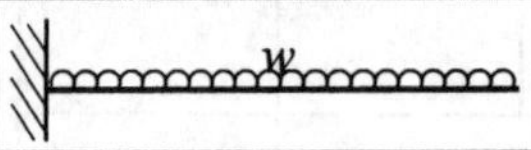 | $\dfrac{wL^3}{6EI}$ | $\dfrac{wL^4}{8EI}$ | $\dfrac{wL^2}{2}$ |
|  | $\dfrac{ML}{2EI}$ | $\dfrac{ML^2}{8EI}$ | $M$ |
| 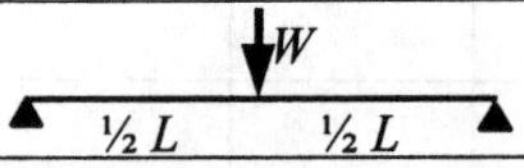 | $\dfrac{WL^2}{16EI}$ | $\dfrac{WL^3}{48EI}$ | $\dfrac{WL}{4}$ |
| 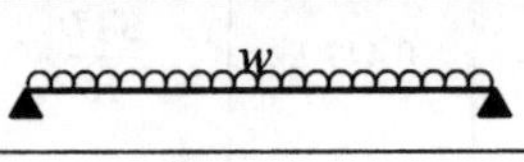 | $\dfrac{wL^3}{24EI}$ | $\dfrac{5wL^4}{384EI}$ | $\dfrac{wL^2}{8}$ |
| 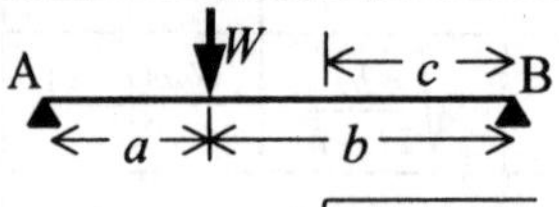<br>$a \le b, \quad c = \sqrt{\tfrac{1}{3}b(L+a)}$ | $\theta_B = \dfrac{Wac^2}{2LEI}$<br>$\theta_A = \dfrac{L+b}{L+a}\,\theta_B$ | $\dfrac{Wac^3}{3LEI}$<br>(at position $c$) | $\dfrac{Wab}{L}$<br>(under load) |

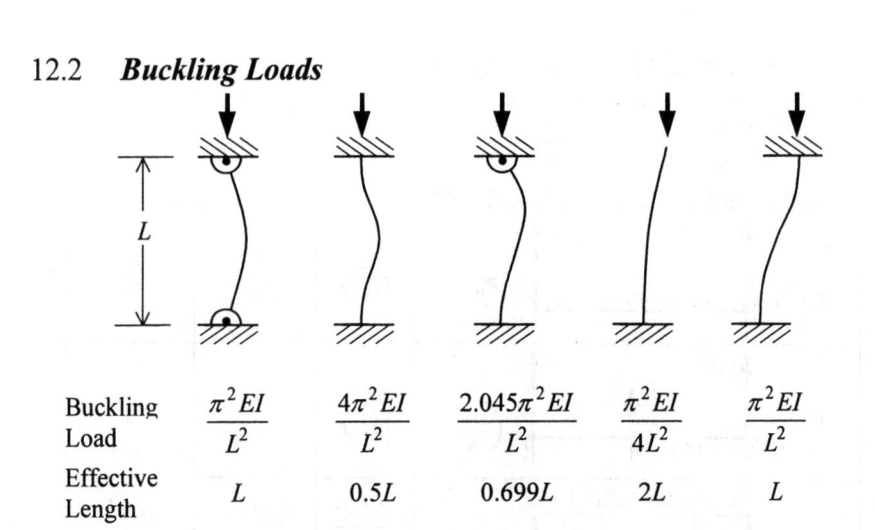

## 12.4 *Fixed End Shear and Moments*

$L$ = overall length, $W$ = point load, $M$ = moment, $w$ = load per unit length

| LH End moment | LH End shear | Diagram | RH End shear | RH End moment | Maximum deflection $\Delta$ | Max deflection position $c$ (from RH end) | Max bending moment (modulus) |
|---|---|---|---|---|---|---|---|
| $\frac{wL^2}{12}$ | $\frac{1}{2}wL$ | 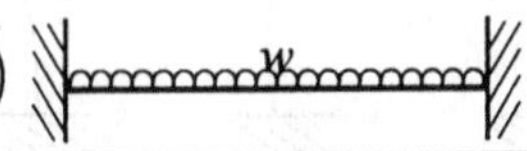 | $\frac{1}{2}wL$ | $\frac{wL^2}{12}$ | $\frac{wL^4}{384EI}$ | $\frac{L}{2}$ | $\frac{wL^2}{12}$ |
| $\frac{WL}{8}$ | $\frac{1}{2}W$ | 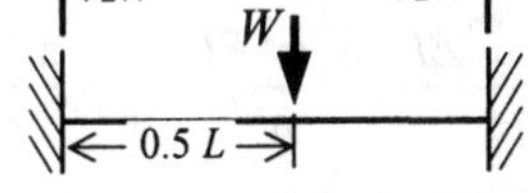 | $\frac{1}{2}W$ | $\frac{WL}{8}$ | $\frac{WL^3}{192EI}$ | $\frac{L}{2}$ | $\frac{WL}{8}$ |
| $\frac{Wab^2}{L^2}$ | $\frac{Wb^2(L+2a)}{L^3}$ | 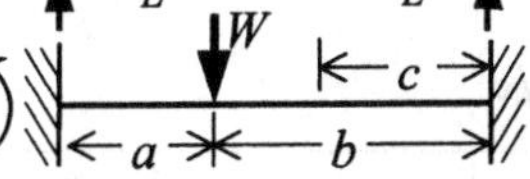 | $\frac{Wa^2(L+2b)}{L^3}$ | $\frac{Wa^2b}{L^2}$ | $\frac{Wa^2bc^2}{6EIL^2}$ | $\frac{2Lb}{L+2b}$ | $\frac{Wab^2}{L^2}$ |
| | | | | | $a \leq b$ | | |
| $\frac{6EI\delta}{L^2}$ | $\frac{12EI\delta}{L^3}$ | 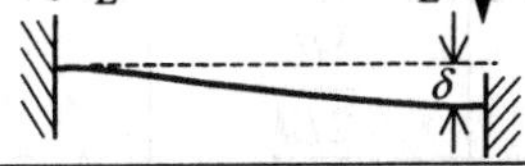 | $\frac{12EI\delta}{L^3}$ | $\frac{6EI\delta}{L^2}$ | $\delta$ | 0 | $\frac{6EI\delta}{L^2}$ |
| $\frac{Mb}{L^2}(2a-b)$ | $\frac{M}{L}$ | 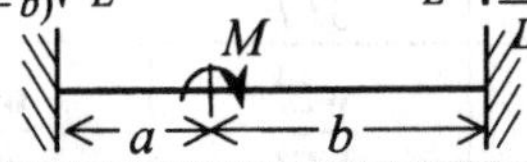 | $\frac{M}{L}$ | $\frac{Ma}{L^2}(2b-a)$ | - | - | $\frac{M}{L^2}a(2b-a)$ |
| $\frac{wL^2}{30}$ | $\frac{3wL}{20}$ | 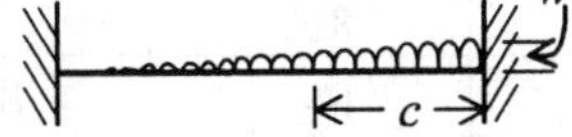 | $\frac{7wL}{20}$ | $\frac{wL^2}{20}$ | $\frac{wL^4}{764EI}$ | $0.475\,L$ | $\frac{wL^2}{20}$ |
| $\frac{3WL}{16}$ | $\frac{11W}{16}$ |  | $\frac{5W}{16}$ | 0 | $\frac{2WL^3}{215EI}$ | $0.447\,L$ | $\frac{3WL}{16}$ |
| $\frac{Wab(L+b)}{2L^2} = M$ | $\frac{Wb+M}{L}$ | 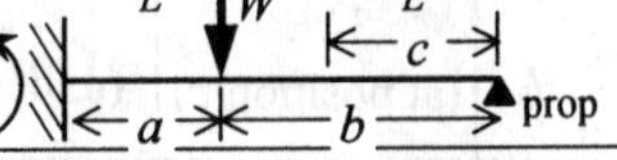 | $\frac{Wa-M}{L}$ | 0 | $\frac{Wa^2bc}{6EIL}$ | $L\sqrt{\frac{b}{2L+b}}$ | $\frac{Wab(L+b)}{2L^2}$ |
| | | | | | $b \geq 0.4142\,L$ | | |
| $\frac{wL^2}{8}$ | $\frac{5wL}{8}$ | 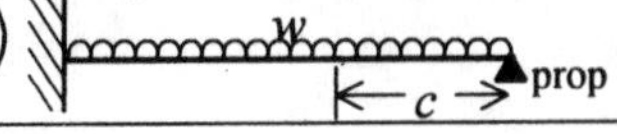 | $\frac{3wL}{8}$ | 0 | $\frac{wL^4}{185EI}$ | $0.442L$ | $\frac{wL^2}{8}$ |

## 12.5 ***Beam stiffness coefficients***

In the following, $F_i$ represent axial or shear forces or bending or torsional couples, according to context; $x_i$ represent linear or rotational displacements as appropriate.

All beam and frame stiffness matrices may be built up from the following orthogonal components for each beam element:

(a) axial:

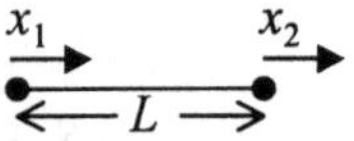


$$\begin{bmatrix} F_1 \\ F_2 \end{bmatrix} = \frac{EA}{L} \begin{bmatrix} 1 & -1 \\ -1 & 1 \end{bmatrix} \begin{bmatrix} x_1 \\ x_2 \end{bmatrix}$$

(b) torsional:

$$\begin{bmatrix} F_1 \\ F_2 \end{bmatrix} = \frac{GJ}{L} \begin{bmatrix} 1 & -1 \\ -1 & 1 \end{bmatrix} \begin{bmatrix} x_1 \\ x_2 \end{bmatrix}$$

(c) plane bending:

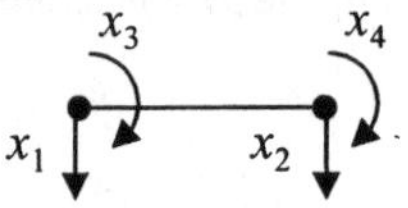


$$\begin{bmatrix} F_1 \\ F_2 \\ F_3 \\ F_4 \end{bmatrix} = \frac{EI}{L^3} \begin{bmatrix} 12 & -12 & 6L & 6L \\ -12 & 12 & -6L & -6L \\ 6L & -6L & 4L^2 & 2L^2 \\ 6L & -6L & 2L^2 & 4L^2 \end{bmatrix} \begin{bmatrix} x_1 \\ x_2 \\ x_3 \\ x_4 \end{bmatrix}$$

Other forms of (c) arise if one end is pinned and vertically constrained:

(i)

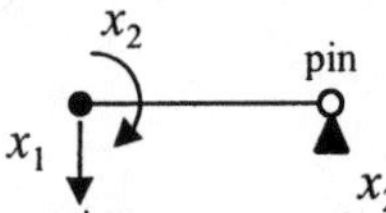


$$\begin{bmatrix} F_1 \\ F_2 \end{bmatrix} = \frac{EI}{L^3} \begin{bmatrix} 3 & 3L \\ 3L & 3L^2 \end{bmatrix} \begin{bmatrix} x_1 \\ x_2 \end{bmatrix}$$

(ii)

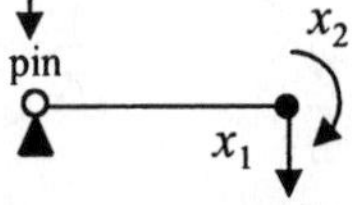


$$\begin{bmatrix} F_1 \\ F_2 \end{bmatrix} = \frac{EI}{L^3} \begin{bmatrix} 3 & -3L \\ -3L & 3L^2 \end{bmatrix} \begin{bmatrix} x_1 \\ x_2 \end{bmatrix}$$

In all cases, if one end is fixed and considered as a reaction, its forces and deflections may be ignored with a corresponding reduction in the stiffness matrix.

The three components (a), (b) and (c) may be combined into larger matrices with zeros in all unspecified positions.

A general plane frame element will have components (a) and (c), and require a 6 × 6 stiffness matrix. A general space frame element will have components (a) and (b), and two components of (c), for two different planes of bending. It will require a 12 × 12 stiffness matrix. In general, the two components of (c) will have different values of $I$.

## 12.6 *Shear*

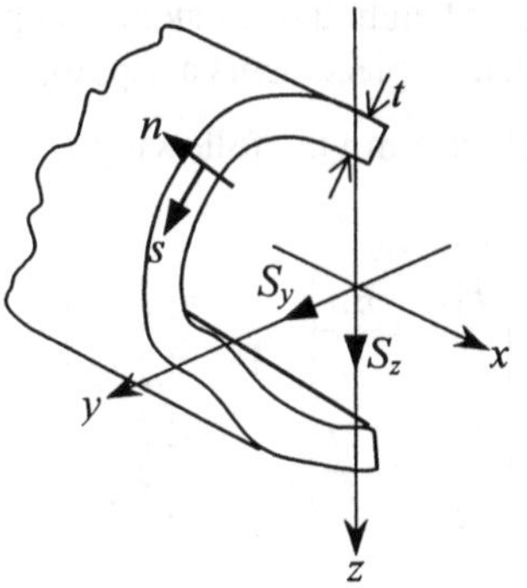


Shear flow per unit length of wall resulting from the *applied* shear forces $S_z$, $S_y$ is

$$q = t\tau_{xs} = \frac{(-S_z)}{I_{yy}I_{zz} - I_{yz}^2}\left(I_{zz}\int_A z\mathrm{d}A - I_{yz}\int_A y\mathrm{d}A\right) + \frac{(-S_y)}{I_{yy}I_{zz} - I_{yz}^2}\left(I_{yy}\int_A y\mathrm{d}A - I_{yz}\int_A z\mathrm{d}A\right)$$

Resultant force for this shear flow acts through the *Shear Centre.*

## 12.7 *Torsion*

For a thin walled closed section,

$$T_x = 2Aq = \frac{4A^2G}{\int_0^S \frac{1}{t}\mathrm{d}s} \times \frac{\theta}{L}$$

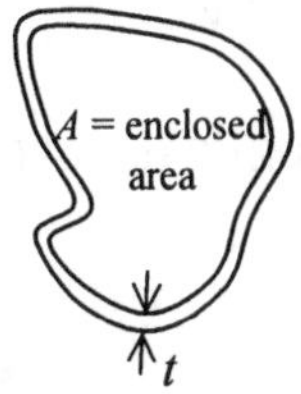


For a thin rectangular section,

$$T_x = \frac{dt^2}{3}\tau_{zx\max} = \frac{dt^3G}{3}\frac{\theta}{L}$$

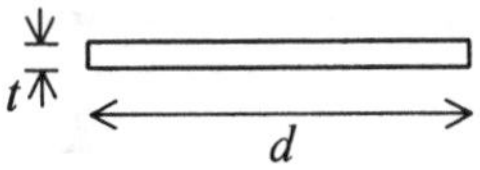


(For torsion of a circular rod or tube, see page 12-1)

## 12.8 *Asymmetric bending*

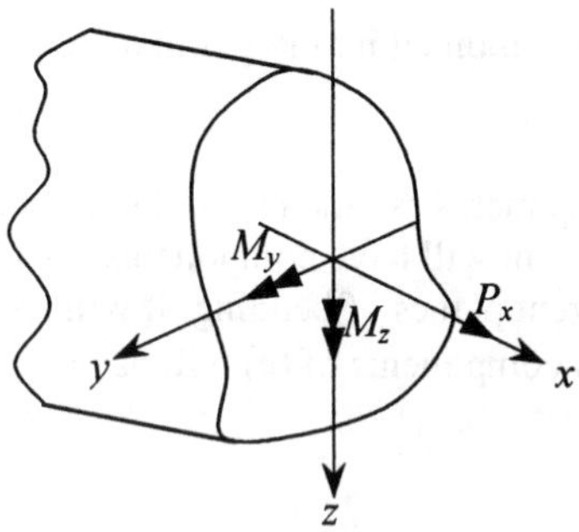


In terms of general axes,

$$\sigma_{xx} = \frac{P_x}{A} + \frac{M_y(zI_{zz} - yI_{yz})}{I_{yy}I_{zz} - I_{yz}^2} - \frac{M_z(yI_{yy} - zI_{yz})}{I_{yy}I_{zz} - I_{yz}^2}$$

When the principal axes $m$, $n$ lie in directions $y$ and $z$, $I_{yz}$ will be zero, and

$$\sigma_{xx} = \frac{P}{A} + \frac{nM_n}{I_{mm}} - \frac{mM_n}{I_{nn}}$$

## 12.9 *Stress and strain transformations*

### Mohr's circle for two-dimensional stress system

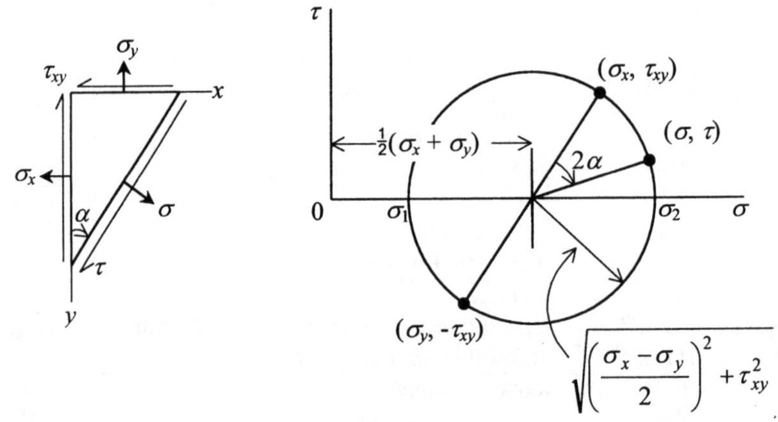


### Mohr's circle for two-dimensional strain

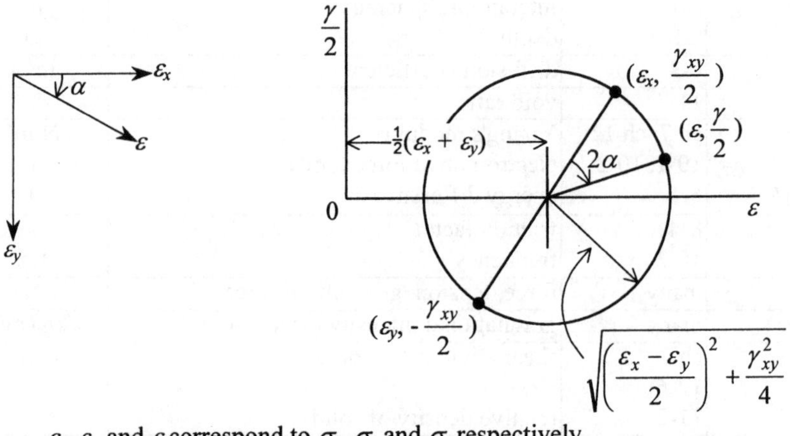


$\varepsilon_x$, $\varepsilon_y$, and $\varepsilon$ correspond to $\sigma_x$, $\sigma_y$ and $\sigma$, respectively.

$\frac{\gamma_{xy}}{2}$ and $\frac{\gamma}{2}$ correspond to $\tau_{xy}$ and $\tau$, respectively.

### Three-dimensional stress system

Principal direct stresses $\sigma_1$, $\sigma_2$ and $\sigma_3$ are equivalent to principal shear stresses $\frac{1}{2}(\sigma_1 - \sigma_2)$, $\frac{1}{2}(\sigma_2 - \sigma_3)$ and $\frac{1}{2}(\sigma_3 - \sigma_1)$.

Strain energy per unit volume:

$$U = \frac{(\sigma_1 + \sigma_2 + \sigma_3)}{18K} + \frac{\left[(\sigma_1 - \sigma_2)^2 + (\sigma_2 - \sigma_3)^2 + (\sigma_3 - \sigma_1)^2\right]}{12G}$$

# 13. Symbols Index

N.B. Symbols with very clear conventional meanings, symbols which are adequately defined where they appear in the text, symbols locally modified with subscripts and symbols which represent constants in a particular context are omitted from this list. Units given are the basic units, without the multipliers which may be used in practice.

| Symbol | Page used | Quantity | SI unit |
|---|---|---|---|
| $a$ | 5-3 | acceleration | $m/s^2$ |
| | 8-8 | velocity of sound | m/s |
| | 10-1 | area | $m^2$ |
| $A$ | 5-1, 5-2 | area | $m^2$ |
| | 6-1 | atomic weight | - |
| $B$ | 10-1 | magnetic flux density | T |
| $c$ | 8-2 | velocity | m/s |
| $c_p$, $c_v$ | ch 7, 8 | specific heat at constant pressure, volume | J/(kg K) |
| $c_v$ | 11-4 | consolidation coefficient | |
| $C$ | 6-2 | concentration | $mol/m^2$ |
| | 8-8 | Chézy coefficient | $m^{1/2}/s$ |
| | 10-1, 10-3 | capacitance | F |
| $d$ | 6-1 | interatomic spacing | m |
| | 8-8 | depth | m |
| $D$ | 6-2 | diffusion coefficient | $m^2/s$ |
| $e$ | 11-2 | void ratio | - |
| $E$ | ch 7, ch 12 | Young's modulus | $N/m^2$ |
| | 10-1, 10-2 | electromotive force (emf) | V |
| $\Delta E$ | 6-1 | energy difference | J |
| $f$ | 8-4 | friction factor | - |
| | 10-2 | frequency | Hz |
| $F$ | many | force, tension, generalised force | N |
| $g$ | many | gravitational intensity (acceleration) | N/kg ($m/s^2$) |
| $G$ | ch 7, 12-1, 12-6 | shear (rigidity) modulus | $N/m^2$ |
| $G_s$ | 11-2 | relative density of solids | - |
| $h$ | 8-3, 8-9 | heat transfer coefficient | $W/(m^2K)$ |
| | 8-5 | bearing clearance | m |
| $h$, $h_{fg}$ | 8-1 | specific enthalpy, of evaporation | J/kg |
| $H$ | ch 8 | enthalpy | J |
| | 8-7 | head | m |
| | 10-1 | magnetic field strength | A/m |
| $i$ | 11-2 | hydraulic gradient | - |
| $I$, $I_{nn}$ | ch 5, ch 12 | second moment of area (about axis *nn*) | $m^4$ |
| | | moment of inertia (about axis *nn*) | kg $m^2$ |
| $I$ | ch 10 | current | A |
| $J$ | 5-4, 12-1 | polar moment of inertia | kg $m^2$ |
| | 6-2 | diffusion flux | $kg/(m^2s)$ |

| Symbol | Page used | Quantity | SI unit |
|---|---|---|---|
| $k, k_{nn}$ | ch 5, 8-4 | radius of gyration (about axis *nn*) | m |
| $k$ | 6-2 | Boltzmann's constant | J/K |
| | ch 7, 8-9 | thermal conductivity | W/(m K) |
| | 8-7 | cavitation number | - |
| $K, K_{IC}$ | 6-3, ch 7 | stress intensity factor | $Nm^{-3/2}$ |
| | ch 7, 12-1, 12-6 | bulk modulus | $N/m^2$ |
| $\Delta K$ | 6-4 | stress intensity range | $Nm^{-3/2}$ |
| $K_p$ | 8-2 | equilibrium constant | $Pa^{1/n}$ |
| $l$ | many | length | m |
| $\ell$ | 10-1 | total length of conductor | m |
| $L$ | many | length | m |
| | 10-1, 10-3 | inductance | H |
| $m$ | many | mass | kg |
| | 8-8 | hydraulic mean depth (radius) | m |
| $M$ | many | mass | kg |
| | 8-1 | molecular weight | - |
| | 6-3, ch 12 | (bending) moment | Nm |
| $n$ | 6-1 | atoms per unit cell | - |
| | 8-2, 8-8 | polytropic index | - |
| | 8-8 | Manning roughness coefficient | - |
| | 8-9, 10-2 | frequency, rotational speed | Hz, $s^{-1}$ |
| | 11-2 | porosity | - |
| $N$ | 5-3 | normal force | N |
| | 6-1 | Avogadro's number | - |
| | 6-1 | quadratic Miller indices | - |
| $N'$ | 10-1, 10-2 | number of turns | - |
| $N_p$ | 6-2 | total number of atoms | - |
| $N_p$ | 8-2 | number of power cycles/unit time | $s^{-1}$ |
| $N_i$ | 6-4 | total cycles to failure at strain amplitude | - |
| | 6-4 | total cycles to failure at stress amplitude | - |
| $p, \Delta p$ | ch 8 | pressure (difference) | Pa |
| $p_v, p_\infty$ | 8-7 | vapour, reference (free stream) pressure | Pa |
| $p_0$ | 8-8 | stagnation pressure | Pa |
| $p$ | 10-2 | no of pole pairs | - |
| $p'$ | 11-3 | average effective stress | $N/m^2$ |
| $P$ | 8-4 | total (stagnation) pressure | Pa |
| $P_x$ | 12-5 | axial force | N |
| $q$ | 11-2 | flow rate | $m^3/s$ |
| | 11-3 | deviator stress | $N/m^2$ |
| $q_b$ | 8-3 | emissive power of a black body | $W/m^2$ |

| Symbol | Page used | Quantity | SI unit |
|---|---|---|---|
| $Q$ | ch 8 | heat | J |
| | ch 8 | volume flow rate, discharge | $m^3/s$ |
| | 6-2 | activation energy | J |
| | 6-3 | (shear) force, load | N |
| | 10-1 | charge | C |
| $R$ | 7-2 | gas constant | J/(kg K) |
| | 8-8 | hydraulic mean radius (depth) | m |
| | ch 10 | resistance | Ω |
| | 12-2 | radius of curvature | m |
| $R_0$ | 6-2 | universal gas constant | J/(kg mol K) |
| $s$ | ch 8 | specific entropy | J/(kg K) |
| | 10-2 | fractional slip | - |
| $s_{fg}$ | 8-3 | specific entropy of evaporation | J/(kg K) |
| $S$ | ch 8 | entropy | J/K |
| $S$, $S_f$, $S_0$ | 8-8 | channel, friction, invert slope | - |
| $S_y$, $S_z$ | 12-5 | shear force | N |
| $T$ | 5-3, 8-5 | torque | Nm |
| | 12-1, 12-5 | | |
| T, $\Delta T$ | many | temperature, temperature difference | K |
| $T_0$ | many | stagnation/reference temperature | K |
| $T_c$ | 7-3 | critical temperature | K |
| $u$ | 5-3 | initial velocity | m/s |
| | 11-2 | pore water pressure | Pa |
| $u$, $u_m$, $u_\infty$ | ch 8 | velocity, free stream (external) velocity | m/s |
| $u_\tau$ | | "friction velocity" | m/s |
| $u$, $u_{fg}$ | 8-1,8-3 | specific internal energy, of evaporation | J/kg |
| $U$ | ch 8 | internal energy | J |
| $v$ | 5-3, 11-2 | velocity | m/s |
| | 8-1, 11-2 | specific volume | $m^3/kg$ |
| | 11-1 | volume | $m^3$ |
| $v_\infty$ | 8-7 | reference (free stream) velocity | m/s |
| $v_s$ | 8-2 | swept volume | $m^3$ |
| $V$ | 6-1 | volume of unit cell | $m^3$ |
| | ch 8 | volume | $m^3$ |
| | ch 10 | voltage | V |
| $w$ | 11-2 | water content | - |
| | ch 12 | load/unit length | N/m |
| $W$ | ch 8 | work | J |
| | 10-4 | power | W |
| | ch 12 | load | N |
| $X_0$ | 10-2 | leakage reactance per phase | Ω |
| $z$ | 8-2, 8-4 | height, head, elevation | m |

| Symbol | Page used | Quantity | SI unit |
|---|---|---|---|
| $Z$ | 8-4 | total head | m |
| $Z, \underline{Z}$ | 10-3 | impedance, complex | Ω |
| $\alpha$ | 6-6, ch 7<br>10-1 | coefficient of thermal linear expansion | $K^{-1}$ |
| $\beta$ | 7-2, 8-9 | coefficient of thermal volume expansion | $K^{-1}$ |
| $\gamma$ | 7-2 | ratio of specific heats | - |
| $\gamma, \gamma_w$ | ch 11 | unit weight of soil, water | $N/m^3$ |
| $\delta, \delta^*$ | 8-5 | boundary layer thickness, displacement | m |
| $\Delta$ | 8-7 | dimensionless diameter | - |
| $\varepsilon, \varepsilon_0, \varepsilon_r$ | 6-1, 10-1 | permittivity, free space, relative | F/m, F/m, - |
| $\varepsilon$ | 8-3 | emissivity | - |
| | 8-5 | eddy viscosity | $Ns/m^2$ |
| $\varepsilon_p$ | 6-4 | plastic straining range | - |
| $\xi$ | 3-13 to 3-15 | critical damping ratio | - |
| $\theta$ | many | angle | rad |
| | 8-5 | boundary layer momentum thickness | m |
| $\dot{\theta}, \ddot{\theta}$ | 5-3 | angular velocity, acceleration | rad/s, $rad/s^2$ |
| $\lambda$ | 8-9 | mean free path | m |
| $\mu$ | 5-3, 5-4 | coefficient of friction | - |
| | 7-2, ch 8 | dynamic viscosity | $Ns/m^2$ |
| $\mu, \mu_0, \mu_r$ | 10-1, 10-2 | permeability, free space, relative | H/m, H/m, - |
| $\nu$ | ch 7, 12-1 | Poisson's ratio | - |
| | ch 8 | kinematic viscosity | $m^2/s$ |
| $\rho$ | many | density | $kg/m^3$ |
| $\rho_0$ | 10-1 | resistivity | Ωm |
| $\sigma$ | 6-3, 6-4,<br>11-2, 12-1 | direct stress | $N/m^2$ |
| | 7-2, 8-9 | surface tension | N/m |
| $\sigma'$ | 11-2, 11-3 | effective stress | $N/m^2$ |
| $\sigma, \sigma_{th}$ | 8-7 | cavitation number (Thoma) | - |
| $\sigma_y, \sigma_f$ | ch 7 | yield (proof), failure (ultimate) stress | $N/m^2$ |
| $\tau, \tau_0$ | 8-4 to 8-6<br>12-1 | shear stress, surface (wall) shear stress | $N/m^2$ |
| $\phi$ | many | angle | rad |
| | 5-3 | angle of friction | |
| $\Phi$ | 10-1, 10-2 | magnetic flux | Wb |
| $\omega, \underline{\omega}$ | ch 5, 8-5 | angular velocity (vector) | rad/s |
| $\omega, \omega_0$ | 3-13, 10-3,<br>3-14 | frequency, natural | rad/s |
| $\omega_s$ | 8-7 | dimensionless specific speed | - |

# Index